MANIFESTOS
ON THE FUTURE OF FOOD & SEED

edited by
vandana shiva

featuring essays by
Carlo Petrini, Michael Pollan, Prince Charles,
Vandana Shiva, Jamey Lionette, and the
International Commission on the Future of Food
and Agriculture

South End Press
Cambridge, Massachusetts
Read. Write. Revolt.

Cover art: *Process* by Nikki McClure
Cover design: Annie Sung
Page design and production: South End Press Collective/Jocelyn Burrell

Library of Congress Cataloging-in-Publication Data
Manifestos on the future of food & seed / edited by Vandana Shiva.
   p. cm.
 Includes index.
 ISBN 978-0-89608-777-4 (alk. paper)
1. Food supply. 2. Food industry and trade. 3. Sustainable agriculture.
I. Shiva, Vandana.

HD9000.6.M35 2007
338.1'9--dc22

2007025404

12 11 10 09 08 07   1 2 3 4 5 6 7

Printed by union labor on acid-free, recycled paper.
Printed in Canada

South End Press ° 7 Brookline Street, #1 ° Cambridge, MA 02139
http://www.southendpress.org

THE MANIFESTOS ✺

BELLY OF THE BEAST $5.99/LB: THE FUTURE OF FOOD IN THE US ✺

Contents

INTRODUCTION

Terra Madre:
A Celebration of Living Economies

vandana shiva

IN A WORLD DOMINATED by fear and fragmentation, dispensability and despair, a magical gathering of food communities—Terra Madre—took place in Turin, Italy, in October 2004. Slow Food, the movement that has put the culture of growing and eating good, healthy, diverse food at the heart of social, political, and economic transformation, brought together 5,000 members from 1,200 food communities in 130 countries. Despite the diversity and differences, everyone was connected: connected through the earth, our mother, Terra Madre; connected through food, the very web of life; connected through our common humanity, which makes a peasant the equal of a prince.

Terra Madre was a gathering of small producers who refuse to disappear in a world where globalization has written off diversity of species and cultures, small producers, local economies, and indigenous knowledge. Not only are small farmers and local food communities refusing to go away, they are determined to shape a future beyond globalization. As Giovanni Alemanno, the Italian minister of agriculture and forestry, said in his introductory speech at Terra Madre:

> What is original and truly revolutionary about Terra Madre is that by selecting the food communities least susceptible to industrial process—hence distinctive for the authenticity and quality of their produce—it attempts to place small-scale food producers at center stage.

Over the past few decades, food production, processing, and distribution has shifted out of the hands of women, small farmers, and small producers and is being monopolized by global corporate giants such as Cargill, Monsanto, Phillip Morris, and Nestlé. Small producers everywhere are being displaced and uprooted by the unfair competition from heavily subsidized agribusiness. The antiglobalization movement has focused on the unfairness of global trade rules that are pushing farmers into debt and suicide. At Terra Madre, small producers gathered not just to curse the darkness of corporate globalization but also to light and keep lit the lamps of small decentralized, biodiverse production.

The vibrant energy of Terra Madre came from the resilience of producers who had continued to save and share their diverse seeds, live their diverse cultures, speak their diverse languages, and celebrate their diverse food traditions. There was a community of dried mango producers; entomophagous women of Ouagadougou, Burkina Faso (women who harvest, process, and sell edible insects); the Baobab community of Atacora, Benin; basil growers; makers of Liguria; nomadic shepherds from India and Kirgbity; sheep breeders from Central Asia; jasmine rice producers from Thailand; and basmati rice producers from India (these last two have both been biopiracy victims of RiceTec). The world of Terra Madre reflected the real world of people—with diversity so dazzling that the eyes and ears were having a feast, while communities communicated with pride, joy, and dignity about their agricultural and food traditions. This was not the world of the World Trade Organization (WTO), where only agribusiness exists; where agricultural trade basically means soya, corn, rice, wheat, and canola; where one company (Monsanto) accounts for 94 percent of the world's genetically modified organisms (GMOs); and where most food grown is not eaten by humans but by billions of captive animals in factory farms. In Terra Madre's world, small farms produce more than industrial farms while using fewer resources; biodiversity protects the health of the soil and the health of people; and quality, taste, and nutrition are the criteria for production and processing, not toxic quantity and superprofits for agribusiness.

Diversity provides the ground for us to turn our food systems around—diversity of crops, diversity of foods, and diversity of cultures. Diversity provides both the resistance to monocultures and the creative alternative. Our strength is our uniqueness and variety, a strength that can be eroded only when we give up on it ourselves.

ANOTHER PARADIGM FOR FOOD

Terra Madre provided an opportunity and platform to articulate another paradigm for food. During the opening ceremony, Carlo Petrini, the founder of Slow Food, called upon everyone to defend the rights, knowledge, and creativity of small producers all over the world. He also called for us to abandon the gap between consumer and producer. "Let us become coproducers," he said. To consume means to destroy. That's why "consumption" was the name given to tuberculosis. In the act of eating, we are already participating in production. By eating organic, we are saying no to toxins and supporting the organic farmer. By rejecting GMOs, we are voting for the rights of small farmers and people's right to information and health. By eating local, we are taking power and profits away from global agribusiness and strengthening our local food community. Eaters are, therefore, also coproducers, both because their relationship with small producers is a critical link in creating a sustainable, just, healthy food system and because we are what we eat. In making food choices, we make choices about who we are.

The industrialization and globalization of our food systems is dividing us: North-South, producer-consumer, rich-poor. The most significant source of our separation and division is the myth of "cheap" food, the myth that industrial food systems produce more food and hence are necessary to end poverty. However, small, biodiverse organic farms have higher output than large industrial monocultures.

As Prince Charles of Great Britain reminded the gathering during the closing ceremony:

> One of the arguments used by the "agricultural industrialists" is that it is only through intensification that we will be able to feed an expanded world population. But even without significant investment, and often in the face of official disapproval, improved organic practices have increased yields and outputs dramatically. A recent UN-FAO study revealed that in Bolivia potato yield went up from 4 to 15 tons per hectare. In Cuba, the vegetable yields of organic urban gardens almost doubled. In Ethiopia, which 20 years ago suffered appalling famine, sweet potato yields went up from 6 to 30 tons per hectare. In Kenya, maize yields increased from 2 1/4 to 9 tons per hectare. And in Pakistan, mango yields have gone up from 7 1/2 to 22 tons per hectare.

During the inaugural ceremony, I drew attention to the fact that globalized, industrialized food is not cheap. It is too costly for the earth, for the farmers, and for our health. The earth can no longer carry the burden of groundwater mining, pesticide pollution, disappearance of species, and destabilization of the climate. Farmers can no longer carry the burden of debt inevitable in industrial farming.

That 150,000 farmers have committed suicide in India is a symptom of the deep crisis in the dominant model of farming and food production. This system is denying the right of food and health both to the 1 billion who are hungry and the 1 billion who suffer from obesity. It is incapable of producing safe, culturally appropriate, tasty, quality food. And it is incapable of producing enough food for all because it is wasteful of land, water, and energy. Corporate agriculture uses ten times more energy than it produces, ten times more water than ecological agriculture. It is thus ten times less efficient. Labor efficiency is also a myth; all the researchers, pesticide producers, genetic engineers, truck drivers, and soldiers engaged in wars over oil are part of the industrial food production system. If all the people involved in nonsustainable food production were counted, the labor efficiency of industrial food would also be lower than that of ecological food. When agriculture becomes like war, and weapons of mass destruction replace internal farm inputs, food becomes nonfood. Trade based on false prices and unfair exchange is not trade; it is exploitation.

Industrial food is cheap not because it is efficient—either in terms of resource or energy efficiency—but because it is supported by subsidies and it externalizes all costs—the wars, the diseases, the environmental destruction, the cultural decay, the social disintegration.

Terra Madre was a celebration of an honest agriculture in which prices do not lie, which does not exploit the earth and the earth's caretakers. Terra Madre was a celebration of our practice of living economies in which we coproduce with the earthworm and the spider, with the mycorrhiza and the fungus. We are all connected in the web of life, and it is food that spins that web. As the ancient *Taitreya Upanishad* tells us: "From food, all creatures are produced.... Beings are born from food, when born they live by food, on being deceased, they enter into food."

In India, we are creating food democracy through freedom farms, freedom villages, and freedom zones. Organic farms free of chemicals and toxins and zones free of corporate—that is, GMOs—and patented seeds are creating a bottom-up democracy of food to counter the top-down food dictatorship.

VOICES FROM TERRA MADRE

Communties of Food[*]

carlo petrini

WHEN WE FIRST came up with the idea of Terra Madre, none of us imagined that as many as 1,200 food communities from 150 countries in every part of the planet would converge together. Founded on sentiment, fraternity, and the rejection of egoism, these food communities—farmers, fishers, breeders, nomads from the Peruvian Andes to the Argentine pampas, from the Amazon jungle to the Chiapas mountains, from Californian vineyards to First Nation reserves, from the shores of the Mediterranean to the seas of northern Europe, from the Balkans to Mongolia, from Africa to Australia—have a strategic importance in designing a new society, a society based on fair trade. Through their labor, food communities bind together the destinies

*October 2004, 2006

of women and men pledging to defend their own traditions, cultures, and crops.

These communities are depositaries of ancient and modern wisdom. They are an important and strategic factor in human nutrition, in the delicate balance between nature and culture that underpins our very existence. Food communities are the expression of human labor: in the sectors of agriculture, animal breeding, fishing, herding, and food processing. These communities practice the fine arts that turn milk into cheese, grapes into wine, and malt and hops into beer; they are also the expression of the earliest human interaction with nature: cooking.

Cooking is language, cooking is identity, cooking is a primary need of all humankind. It is culinary skill, manual dexterity, and the ability to capture the right mix of flavor and spiciness that makes eating pleasurable. This pleasure has never been and will never be the privilege of the few. It is one of the physiological prerogatives of all us, a sign of humankind's serene relationship with nature and life.

All around the world, no one food culture is more important than another. Every single one precisely expresses a profound identity and its language through food. We have to respect these diversities. We have to be grateful to the art and skill of women and men capable of producing foods as simple as they are outstanding—the fruit of an ability to exploit and make do with the sometimes scant resources that Mother Nature has made available to us. Feijoada in Brasil, couscous in the Mediterranean, tamales in Latin America, pakora in India, fufu in Africa, dried reindeer meat in Lapland, pasta in Italy—all foodstuffs representative of the great wisdom of humankind, of subsistence economies, and of the never-ending fight against hunger.

Terra Madre reveals not only evidence of this wisdom but also the environmental, social, and economic problems that affect our daily labor. This extraordinary knowledge and know-how must not be threatened by the logic of productivity, by the manipulation of genes, by the profit motive of a privileged few, by a lack of respect for the environment, by the exploitation of workers.

The battle that we are waging to defend the biodiversity of the planet—from vegetable species to animal breeds—is a battle for civilization. The right to own land and seeds is a sacrosanct right for all the world's vegetable growers. The pesticide- and genetically modified organism (GMO)-multinationals are implementing policies incompatible with the environment, that stress Mother Earth, that humble the food sovereignty of peoples, and that jeopardize the freedom of farmers and growers.

Starting with the first Terra Madre, we have prepared the ground: we have spread good fertilizer, we have plowed, we have broken up the soil, and now we are ready to plant our seeds. These are the seeds that will make the network of Terra Madre a tangible and believable thing. And now, all that remains is to ask, What is the seed? What is the distinctive sign of this extraordinary session, with citizens, shepherds, nomads, and fisher persons that have come to Turin from 150 different countries? What is the seed that we have to plant? I've given it some thought, and I think that the strong seed of Terra Madre is the practice of the local economy: the local economy that all of us want to bring about in the villages and in the countryside. The local economy that you all know so well is based on three principles: the principle of solidarity, the principle of support, and the principle of subsistence.

Solidarity and subsistence are the strength of the local community, and it is in this way that people can produce food. In producing food locally, people adopt the habits that shorten the distance between the producer and the consumer, that contribute to the well-being of the community, that help those who work in the fields to prosper, that give health, that give beauty to their own land. This local economy is in perfect harmony with nature. Look! The communities are primarily a place, a place and a people: the people of a certain place and their local economy are extraordinarily compatible with a philosophy of sustainable development.

The land is our common home, and our common home must be governed by an honest economy; it must be governed by a natural economy. What governs this earth, on the contrary, with force and often with arrogance, is the iron hand of the market economy; we must not confuse the market economy with the local economy. The distortions of the same market economy that at the beginning had beneficial effects for communities are evident to everyone today—the world's resources are not infinite. And yet, you can't apply a law to regulate the exponential extraction of resources because the market currently demands hyperproductivity. This generates an unsustainable food situation that is beyond madness. The Food and Agriculture Organization (FAO) data tells us that we produce food for 12 billion people, when there are only 6.3 billion people living. Meanwhile, 800 million suffer· from malnutrition and hunger, 1.7 billion suffer from obesity, and the rate of diabetes is growing exponentially along with cardiovascular diseases caused by malnutrition. It's madness! It's madness to continue to ask more from the land. This predation of resources, this logic in which consumption must be fast, is taking us from a point of abundance and waste to a terminal point.

The world's food crises are today a fact of life, visible to everyone. We hear daily about water shortages, excessive use of chemical fertilizers, infertile soils, the loss of biodiversity, global warming—the entire ecosystem threatened. People still call it silly, but already the thing is so evident, so strong, that the media can no longer avoid this reality. Communities of people, religious communities, political communities, even the most sensible politicians have already made it clear that this food crisis is dramatic!

The unsustainablity of this economy is becoming clear enough; what is not clear is the degree of our own complicity, our own responsibility as individual consumers in this so-called developed world. To what degree are we in the so-called developed world responsible for, complicit in, an unstoppable consumption? Because of this, it will be difficult for us in the so-called developed world to become the standard-bearers in combating the economy of the market. We are accomplices, we are participants, and citizens living in the so-called underdeveloped world must show us the way, the way toward an economy that relocalizes consumption and agricultural production. What's needed is a return to agricultural production in every single state, and with this, instead of conflict, we must work to build true, frank, and sincere support for the local economy. Food communities validate these local economies' right to exist. However, the damage wrought by the many authors of the market economy is not simply the disaster being created; it is also the ridicule, the sneering, the derision of the local economy. They declare that local economies are not scientific. Supporters of the market economy ask poison questions: "What do you care about these local economies that have no future?"

Here, there is much work to be done, because the vision of local economy is not an archaic one, but an extremely modern vision. The local economy is the only one that allows for the realization of what is becoming an oxymoron: sustainable development. If we want to bring about sustainable development, we must reinforce the elements of the local economy and recognize how much creativity there is in making this local economy. In America, in green California, you bring about this local farming economy: constructing farmers' markets that bring together producers and consumers, giving energy back to organic agriculture. Farmers of India bring about an economy that fights the predominance of seeds and reinforces the agricultural economy. Our own Italy, from the moment she chose to support regional production, from the moment she chose to support methodical modes of production that strengthen the fabric of the tourism associated with our beautiful country, she too brings about, consciously or unconsciously, a strong local economy, to such an extent that the market economy now copies the good ideas of typical products and seeks also to take them away.

We must have the strength to give this economy back to the citizens, because food must be good, clean, and fair. Good, clean, and fair. Good, absolutely good—it is not that we are condemned to eat badly! Even during the historic troubles of this country, many of my gastronome friends speak of the "Italian memory." The Italian gastronomic memory has a name: hunger. Built into this memory is the wisdom of many women, which, along with a subsistence economy, brought about some masterpieces, very simple, but good. Clean, because one cannot produce nourishment by straining ecosystems, ruining the air, and destroying biodiversity. Fair, because the citizen must be paid; if we want the young people to stay and return to the land here in our countries they must have dignity and fulfillment. They must be valued. It is inconceivable that a civilized nation could enslave the workers of other nations to produce tomatoes. It is inconceivable that a civilized country can encourage organic economies like that of green California at the same time that it reduces many Mexican farmers to slavery.

So *good, clean,* and *fair* are three adjectives that farmers must offer to the consumers, whom I would like to call coproducers, in an effort to change this system that is turning into a big mistake. This is the meaning of the international network; this is what virtuous globalization means. We are planting the seeds of a virtuous globalization, and this virtuous globalization must have the strength to claim the collective rights that make up the economy and that bring about a new frontier of rights. The French Revolution strengthened in everyone the knowledge that we have some individual rights; time and history call us to collective rights. The right to water, the right to clean air, the right of women to remain in the systems of agricultural production, the right to defense of biodiversity, the right to peace.

As we concluded the second Terra Madre, we took the opportunity to think about the tangible prospects of this extraordinary meeting and reflect upon the major decisions that we'll have to take in the coming years. To do this, it's important for us to establish who we really are. When we coined the term *food community* two years ago, we were aware that it embodied a very precise concept. Celebrating food as a central, primary element in our lives seems an obvious thing to do, since without food no living things would exist. In the course of time, alas, food has moved from the center, and gastronomy and the culinary art have been reduced to mere folklore with no thought for the true roots of food culture: roots that are the essence of life, roots that make us part of the circle of life itself. We must reinstate the centrality of food.

For this reason, the staging of Terra Madre alongside the Salone del Gusto is charged with significance. The Salone del Gusto is, arguably, the world's largest showcase of agricultural and food products. The Salone del Gusto and Terra Madre aren't taking place simultaneously for reasons of organizational convenience. We are making a major cultural choice. Food quality depends on consumers who respect agricultural labor and educate their senses; thus they are becoming precious allies for producers.

Ten years ago, 75 percent of the exhibitors at the Salone were traders and only 25 percent were producers. Today the proportions are reversed. In many quarters it was thought that the choice to include an ethical, environmentalist option—including the politics and culture of Terra Madre—might somehow reduce attendance at the Salone del Gusto, but exactly the opposite has happened. What does this mean? It means that, even in civil society, the need is felt for a close relationship with production, for coming to terms with others, for finding out more about others, and for developing food culture.

The special chemistry that made this great meeting a success brought 200,000 people from all over the world to the Salone and 6,500 delegates to Terra Madre to debate the fate of the earth, the right to water, the right to clean air, and the right to own seeds. This is the most important element to understand and to be developed in our own communities. Maybe it isn't utopian to hope that here at this meeting of ours we can lay the basis for a food community whose members, despite their geographical distance from each other, can keep in touch and enrich each other through intelligent discussion. And in this way, many of us will feel less alone, proud of our work, and ready to develop the self-esteem that generates well-being and happiness.

The network is brought about also by alliances, and it is for this reason that Terra Madre has cooks and university professors present. The cooks, because it is they, working culturally and with culture, who transform the products of nature. Farmers need cooks and cooks need good farmers, as well as university professors. It is a marvelous thing that among this public assembly of farmers there are 250 university professors from all over the world; it's a marvelous thing because we must safeguard traditional knowledge, which, as the great intellectual Claude Levi-Strauss has said, has been at risk of disappearing throughout the last 30 years. We must safeguard cultural knowledge, wisdom, the traditions of the rural world, and to do this we must forge an alliance with universities and an alliance with official science. We cannot afford more misunderstandings between science and traditional knowledge, and with more dialogue and collaboration, you will see that many problems will be resolved—we will resolve the problem of GMOs if science will speak with the farmers.

I believe that never as in this moment in time have consumers shared a common destiny. The safeguarding of our food heritage is a mutual obligation and as such can only be achieved by new ways of sharing. All the producers and consumers who, like the food communities, conduct themselves sustainably at an economic, ecologic, and social level are committed to fueling the creative force behind every human identity: exchange.

Only if consumers become coproducers and fully grasp the fact that production is being threatened, and only if producers assume the burden of quality—ensuring food safety, sustainability, pleasure, and human rights—can we leave this difficult moment behind us.

These values and this intuition are all embracing, as valid in the North of the world as they are in the South. They spur us to become an active part in the fight against the planetary scourge of malnutrition and hunger.

Producers of culture have strong relationships with their communities and also with Mother Earth. We must seek out alliances in every community, as we have done here. In every country, we must ask official science to liaise with the traditional know-how of farmers, ask cooks and chefs to connect with producers, make sure rural classes are not isolated and become active protagonists, because the defense of this planet is in our hands, and this is so simply because only you can boost local economies as opposed to the market economy. It's not true that the local economy is obsolete and archaic: the local economy is strong and incredibly modern.

Economists and politicians must acknowledge that local economies are much more profound than we might think; that if the market economy did, to a certain extent, produce benefits at a given moment in time, today, thanks to its logic of expecting more and more of the earth, it is wringing disaster not only on the environment but also on human relationships. We are no longer citizens but consumers, no longer producers but people used to create consumer objects. We don't live in a harmonious relationship with creation: we produce too much and then waste what we produce. Local economies, even when it's a subsistence economy—especially when it's a subsistence economy—serve more to defend against hunger than the market economy, and thanks to them, food has resumed a starring role.

Food communities instruct us on how to use the term *community* in the future. The community, you see, moves on rocky ground. Whereas society at large is a structured and hierarchical system, the community is either structured in a rudimentary way or not structured at all. And this is our strength! Let no one think that an association or a party has been formed at Terra Madre. Nothing of the sort! The dialogue or, dare I say, the dialectic between organized, structured society, represented by institutions, parties, associations, and religions, and democratic, even anarchic communities that live in uncertainty—this dialectic between community and society is a new source of fecundity for us.

Let no one have ideas about structuring Terra Madre. Terra Madre is, by its very nature, a free entity, and this enhances the space available to us. I would even go as far as to say that we enjoy ourselves more because in this way we can use our imaginations, be inventive, have courage, travel new roads, think in terms of mutual care and aid (a key element for the community), live for others, translate rights into duties and think about the well-being of all. Here among us, unlike in structured society, there are neither competitors nor consumer objects. Ours is a participatory, "livable" life and, as such, contains an element of uncertainty. Maybe one day we'll lose control of Terra Madre, but it doesn't matter. So far it's gone very well indeed, because it's bound by a cement that politics can't weaken: the cement of affective intelligence. At last, this world of ours, so full of rational intelligence, can boast a get-together that is affective and fraternal. Thanks to this, we don't need to be structured: we can be free and enjoy ourselves and, at once, create a future for ourselves. Thanks to this, we can put into practice the words of a great intellectual: namely, that wisdom is achieved through "no power, a little understanding, a little knowledge and lots and lots of flavor." And who more than us can give the earth flavor? The flavor of food! The joy of food!

Agriculture: The Most Important of Humanity's Productive Activities*

p r i n c e   c h a r l e s

THE FUTURE OF small-scale agriculture and of artisan food producers throughout the world is a vitally important discussion.

The fact that no fewer than 5,000 food producers have gathered in Terra Madre, under the Slow Food banner is a small but significant challenge to the massed forces of globalization, the industrialization of agriculture, and the homogenization of food—which seem somehow to have invaded almost all areas of our life today.

I have always believed that agriculture is not only the oldest but also the most important of humanity's productive activities. It is the engine of rural employment and the foundation stone of culture, even of civilization itself. And this is not just some romantic vision of the past: today some 60

*October 2004

24

percent of the 4 billion people living in developing countries are still working on the land.

So when I read "visions," such as that for the Indian state of Andhra Pradesh, based on transforming traditional, local agricultural economies into "powerhouses" of technological agriculture, based around monoculture, artificial fertilizers, pesticides, and genetically modified organisms (GMOs), my heart sinks. The missing ingredient in these great plans is always sustainable livelihoods. Its absence increases the existing, awful drift towards degraded, dysfunctional, and unmanageable cities.

The one resource the developing world has in abundance is people, so why are we promoting systems of agriculture that negate this advantage and seem bound to contribute directly to further human misery and indignity?

It is a sobering thought that almost all of the next 1 billion of net global population growth (over the next 12 to 15 years) will take place in urban slums. In one slum alone more than 800,000 people, half of them under the age of 15, already live illegally in less than 4 square kilometers of the city. Even more sobering is the thought: what will these conditions breed for the future? Hopelessness, crime, extremism, terrorism? Who will deal with these chickens when they come home to roost on a globalized perch?

Despite the best intentions of many, we have to face up to the fact that often the consequence of globalization is greater unsustainability. It is all very well talking meaningfully of the need for "globalization with a human face," but the reality is frequently somewhat different. Left to its own devices, I fear that globalization will—ironically—sow seeds of ever-greater poverty, disease, and hunger in the cities and lead to the loss of viable, self-sufficient rural populations. I don't think anyone would claim to have many answers, technological or otherwise, about what could possibly be done to reverse this process. The 800,000 people in the slum I mentioned earlier are not simply going to head back to the land overnight. But, surely, the first step to finding solutions is being willing to face up to both the causes and the scale of the problem—and this requires the globalization of responsibility.

I have a feeling that by now it may be quite well-known that I am inclined to doubt whether genetically modified (GM) food, for instance, will be—on balance—a contribution to the greater good of humanity. I am not simply being dogmatic. I believe it is both legitimate and important to ask whether some people's faith in the potential of this and other new technologies is a product of wishful thinking or of the hype generated by people with vested interests. We need to ask, in the long term, are these methods really going to solve humankind's problems or just create new ones? And how will we regulate them effectively? There are a great many examples of earlier, well-meaning attempts to control pests or improve the environment which have gone drastically wrong. And I'm simply not convinced that we have absorbed the lesson, which is that manipulating nature is, at best, an uncertain business.

Even if we discount the potential for disaster, there is still the question of whether this is the right direction to take. If all the money invested in agricultural biotechnology over the last 15 years had been invested in developing and disseminating genuinely sustainable techniques—those that work with, rather than against, the grain of nature—I believe that we would have seen extraordinary, and genuinely sustainable, progress.

The problem, perhaps, is that techniques such as inter-cropping, agroforestry, green manuring, composting, and biological pest control offer less prospect of commercial gain to those who have money to invest. The hundreds of millions of people who would gain are the much-derided practitioners of so-called peasant agriculture, who have very little money, but who are the long-term guardians of biodiversity.

One of the arguments used by the "agricultural industrialists" is that it is only through intensification that we will be able to feed an expanded world population. But even without significant investment, and often in the face of official disapproval, improved organic practices have increased yields and outputs dramatically. A recent United Nations–Food and Agriculture Organization (UN-FAO) study revealed that in Bolivia potato yields went up from 4 to 15 tons per hectare. In Cuba, the vegetable yields of organic urban gardens almost doubled. In Ethiopia, which 20 years ago suffered appalling famine, sweet potato yields went up from 6 to 30 tons per hectare. In Kenya, maize yields increased from $2\frac{1}{4}$ to 9 tons per hectare. And in Pakistan, mango yields have gone up from 7½ to 22 tons per hectare.

Imposing industrial farming systems on traditional agricultural economies is actively destroying both biological and social capital and eliminating the cultural identity which has its roots in working on the land. It is also fueling the frightening acceleration of urbanization throughout the world and removing large parts of humanity from meaningful contact with nature and the food that they eat.

So this "flight from the land" is happening in both developed and developing countries.

Unfortunately, these trends toward urbanization are almost inevitable while societies throughout the world continue to put a low valuation on their food, denigrate food to the status of fuel, and abandon any loyalty to their local and indigenous farmers.

But there is another consequence too. There is now a growing body of evidence that suggests that in the so-called developed world we are in the process of creating a nutritionally impoverished underclass—a generation which has grown up on highly processed fast food from intensive agriculture and for whom the future looks particularly bleak, both from a social and a health standpoint.

Fast food, as Eric Schlosser has pointed out in his brilliant book *Fast Food Nation*, is a recent phenomenon. The extraordinary centralization and industrialization of our food system has occurred over as little as 20 years. Fast food may appear to be cheap food, and in the literal sense it often is. But that is because huge social and environmental costs are being excluded from the calculations. Any analysis of the real costs would have to look at such things as the rise in food-borne illnesses, the advent of new pathogens such as E. coli 0157, antibiotic resistance from the overuse of drugs in animal feed, extensive water pollution from intensive agricultural systems, and many other factors. These costs are not reflected in the price of fast food, but that doesn't mean that our society isn't paying them.

So perhaps, having said all this, you can begin to see why I am such an admirer of the Slow Food Movement and of all the hardworking, indomitably independent people, all over the world, who are part of it.

Only a few years ago, it would have been impossible to imagine that so many people across the world who are either directly involved in small-scale artisan food production, or are interested in consuming the fruits of such labors, would gather together as they have at Terra Madre. This, of course, is a great tribute to the unceasing energy of Carlo Petrini.

Slow food is traditional food. It is also local—and local cuisine is one of the most important ways we identify with the place and region where we live. It is the same with the buildings in our towns, cities, and villages. Well-designed places and buildings that relate to locality and landscape and that put people before cars enhance a sense of community and rootedness. All these things are connected. We no more want to live in anonymous concrete blocks that are just like anywhere else in the world than we want to eat anonymous junk food which can be bought anywhere. At the end of the day, values such as sustainability, community, health, and taste are more important than pure convenience. We need to have distinctive and varied places to live and distinctive and varied food to eat in order to retain our sanity, if nothing else.

The Slow Food Movement is about celebrating the culture of food and about sharing the extraordinary knowledge—developed over millennia—of the traditions involved with quality food production. So it is important to ask how these ideals can be promoted more widely, particularly when we are faced with remorseless pressure to operate on a larger and ever more impersonal scale.

I believe the food communities gathered at Terra Madre are in a better position to answer that question than I, but, for what it's worth, I do believe that simply coming together and sharing ideas, and above all joining the international Slow Food Movement and to create, by the extraordinary process of cross-fertilization and invigoration which takes place at gatherings like Terra Madre, an ever more influential and powerful association that cannot be so easily ignored, the answers will emerge organically. As the old saying goes, there is safety in numbers, and people tend to listen to organizations with a very large membership.

On this theme it does seem to me that the other great food movement with which I am associated, the organic movement, has so much in common with the Slow Food Movement, and this communality of purpose and direction ought to be a source of co-operation and, also of course, celebration! So I do hope that we may see ever closer links between these two important movements.

And the importance of this movement cannot be overstated. That is, after all, why I am here—to try and help draw attention to the fact that in certain circumstances "small will always be beautiful," and to remind people, as John Ruskin in the 19th century did, back in England, that "industry without art is brutality." After all, the food you produce is far more than just food, for it represents an entire culture—the culture of the family farm. It represents the ancient tapestry of rural life: the dedicated animal husbandry, the struggle with the natural elements, the love of landscape, the childhood memories, the knowledge and wisdom learnt from parents and grandparents, the intimate understanding of local climate and conditions, and the hopes and fears of succeeding generations.

Farmer, Chef, Storyteller: Building New Food Chains[*]

michael pollan

IT WAS A PRIVILEGE to address the people gathered at Terra Madre. To look out from the stage at not only the wondrous diversity of people gathered there but also the equally wondrous diversity of species represented—the plants and the animals and the fungi being cared for and defended. The farmers and herders and foragers and fishers and beekeepers that gathered at Terra Madre form a kind of representative body—a worldwide parliament of species. Each person came to speak for themselves and their communities, but also to speak for the sheep and the honeybees, the belted Galloways and the Ossabow pigs, the chili peppers and the chicories— for the biodiversity of the wild and domesticated species that feed us and clothe us and help hold the world steady.

---

[*] October 2006

That biodiversity is endangered today, by a handful of other species grown in vast monoculture. I'm speaking of maize and soy, about the broad-breasted white turkey and the Cornish cross chicken—the imperial species that are the building blocks of our fast-food world. Of the tens of thousands of food species nature offers humankind, we are relying on a dwindling few: a mere eight crops now supply three-quarters of the world's food.

Farmers stand on the front lines of the struggle to defend the Many against the One. Farmers stand for the specific: the plant adapted to the soils of a particular place, the indispensable ingredient of the traditional dish, the unforgettable flavor we count among the wonders of this world.

But it's not enough simply to nurture this genetic diversity on our farms. No species is an island; it only has meaning—has *life*—in the context of a food chain. To keep the genes in the world, the farmer whole, and the eater healthy and delighted, there needs to be a food economy linking the species and the farmer and the eaters.

This is where we come in—the storytellers and the chefs. If monoculture on our plates leads to monoculture in our fields, the opposite is true as well. Thanks to the work of people like Alice Waters, chefs—formerly servants of the rich—have become advocates for the farmers and foragers, shining the bright light of their glamour on the work they do and on the plants and animals they care for. It is one thing to produce a wonderful pig or salad green, but unless it is cooked well and eaten with pleasure, it will soon vanish from the marketplace and the fields. In the US today, it is the chefs who are leading us forward, rebuilding food economies by cultivating the tastes that save the species.

I am not a chef. I am another kind storyteller. I follow and write about food chains, both the healthy ones and the sick, so people can know where their food comes from. It is a job that didn't need doing 50 years ago, before the era of industrial food. That is how I define industrial food: "food that is incomprehensible without the help of a journalist." We have far too much of this food today. Yet when people are shown where their food comes from, and can see how the plants and animals live, they make good choices, vote with their forks for the Many and not the One—for farmers and their plants and animals.

So here then is our common work—to speak for the species, who are saying, oddly enough, "Eat me." By eating them we support the farmer, so that the farm may sustain them. To speak of the value of plants or animals or farmers or chefs or eaters in isolation makes no sense. We need to remember what Sir Albert Howard taught us a half century ago, that what's needed is to treat "the whole problem of health in soil, plant, animal and man as one great subject." This is our work: to re-create and defend the food chains that link us—soil, plant, animal, and eater.

For the Freedom of Food*

vandana shiva

FELLOW EARTH CITIZENS, children of Terra Madre, I'm sure all of you feel like I do that we are creating another world. We are creating a world beyond the Washington consensus, which we should call the Washington fiction. A fiction that say that $3 trillion of fictitious money moving around the world is real wealth. A fiction that assumes that creating war undemocratically is democracy.

In India we deeply believe that this amazing universe, this amazing planet, this amazing earth is connected through the web of food, the web of life. Food—everything is food, and everything that eats that food is someone else's food. That's what connects us. We are food: we eat food, we are made of

---

* October 2006

food, and our first identity, our first wealth, our first health, comes from the making, creating, giving of good food. In India we have an Upanishad that says, "If you give bad food you sin." The highest karma is the production of food in abundance and the giving of good food in generosity.

Terra Madre is the birth of a new freedom movement. Nothing should be able to push human life to indignity, degradation, and extinction. We *are* being pushed into, and in many places living in, a food fascism. Food has become the place for fascism to act. This fascism is seen where the seed is patented and turned into the monopoly property of a handful of corporations—95 percent of genetically modified (GM) seeds are controlled by one corporation, Monsanto. Monsanto then uses the fictitious democracy that created the World Trade Organization (WTO) and the financial conditionalities of the World Bank and the International Monetary Fund (IMF) to force us to give up our seed freedoms, to give up our biodiversity, and to deny the richness of our resources—reducing us to biodiversity serfs.

In India, as the new GM seeds have moved in and as the corporations have started to control the seed supply, hundreds of thousands of farmers have become indebted and ended their lives with suicides. More than 150,000 Indian farmers have been driven to suicide. Monsanto's profits are becoming more valued than human life. We must change that.

But food fascism isn't just in the seed; it's in the methods of production as well. You cannot *not* use GM seeds. I was told by a farmer from Germany that a potato seed called "Linda" is being *banned* just because companies can't make profits from it anymore. Instruments of seed registration, licensing, seed replacement, patenting—there are all kinds of new fascist rules. Europe made the choice to be genetically modified organisms (GMOs)-free, and the WTO was used to tell Europe, "You will have to grow and eat this rot." But the WTO itself, as we know, is dying; it's in intensive care. Russia has said it won't join. We need to use this moment of WTO weakness to tell that fictitious world capital, "Your immoral rule—whether it is farmers being prevented from growing their crops or distributing their seed—is over." And we need to look deeply at the issue of food safety and how it is being used. Take the avian flu: it is identified with wild birds and free-range birds, but that is not where it started. These birds were the victims of a disease that emerged from factory farms. And yet instead of addressing the breeding ground of the disease, we have these people around the world in moon suits, going out and grabbing chickens from women's backyards to kill them. This is another element of food fascism—the fear of the small, the decentralized, the local, the free.

In fact, I would say fascism is about fear of freedom. And we are about love of freedom—passionate, deep, uncompromising love of freedom—the self-organized freedom that Terra Madre is about.

ℒ

TODAY BRASIL HAS BECOME the biggest producer of genetically engineered soya beans. We need to tell President Luis Inácio Lula da Silva: "Stop the destruction of the Amazon, and stop converting your country into the cutting edge of food fascism." What we have been building in Terra Madre is unique. It's unique because it is the defense of the local through a global alliance. It is a defense of diversity through a joining together. And I think one of the self-organized contributions that has propelled some of this process has come from the International Commission on the Future of Food. I want to thank my copresident Claudio Martini, the president of the Region of Tuscany, who took the initiative to launch this commission four years ago. It was through the discussions of the commission, of which Carlo Petrini is a member, that we realized, that if the next step of food freedom was to be taken, eaters of food and producers of food, ecological movements, and movements of gastronomy would have to come together. And that is how the idea of Terra Madre as a sister event of the Salone del Gusto came to be. Terra Madre is where we empower ourselves, go back with new ideas. The commission's new report, the "Manifesto on the Future of Seed," is, I believe, the manifesto of Terra Madre.

I want to share with you its principles and, I hope you will all join this movement for seed freedom and food freedom and make it your own. Like the earth itself, Terra Madre does not discriminate between a desert and a wetland, a mountain and a river valley. Every place is a hospitable place, every plant is a useful plant, every producer is a creative producer. It is that diversity we celebrate in the "Manifesto on the Future of Food." We need to celebrate and protect the diversity of seed, of agricultural systems, of producer/consumer relationships, so we are all not driven into one Wal-Mart model of shopping in prisons as we grow our food in prisons. We celebrate and protect the diversity of our cultures, the diversity of innovation and knowledge. Terra Madre is also a source of freedom, and that is what is so beautiful about Terra Madre as an event. Instead of laws being written behind closed doors, here, in openness, in dialogue, we create a law of the seed based on freedom. Here you will not find three corporations sitting with three governments and telling the world that from now on seeds will be the intellectual property of Monsanto and that trade-related intellectual property rights agreements will govern the world.

Instead, in open dialogue, we proclaim freedom. Freedom of the seed; freedom of the farmers to save seed; freedom of farmers to breed new varieties; freedom from privatization, patenting, and biopiracy; freedom of farmers to exchange and trade seeds—because seed is a commons, meant to be exchanged, meant to be shared; freedom of access to open-source seed, seed that can be reproduced and regenerated; freedom from genetic contamination and GMOs—which means GMO-free zones in agriculture at the regional level, the national level, the *earth level*. That's where we need to move.

But the most important freedom of seed is the freedom

to reproduce—that's what *seed* means. Seeds are the embodiment of the future, the unfolding of life, the potential to keep reproducing, and yet new technologies, like the "terminator technology," like hybrid seeds, are designed to prevent seed from giving rise to seed. The freedom of seeds to reproduce means we will not accept terminator seeds and sterile seeds, which cannot grow.

The seeds of slavery have been bred to respond to chemicals. They have been bred for the convenience of giant machines that need huge amounts of oil. They have been bred for corporate profits. The seeds of the future will be bred by the food communities and the scientists gathered at Terra Madre, and those whom we will work with, and by women— because women are the keepers of the seed. We will breed them to eliminate toxic inputs, to eliminate dependence on fossil fuels and the emission of greenhouse gases. We will breed seeds not for uniformity and monocultures, but for diversity and, most important, for freedom. That's what we are sowing—the seeds of freedom. I hope all of you will join this manifesto—will you?

THE MANIFESTOS

Manifesto on the Future of Food

THIS MANIFESTO IS the outcome of a joint effort among participants in the meetings of the International Commission on the Future of Food, held in late 2002 and early 2003 in Tuscany, Italy. The government of the Region of Tuscany actively participated in and supported the commission's work.

The manifesto is intended as a synthesis of the work and the ideas espoused by hundreds of organizations around the world and thousands of individuals actively seeking to reverse the dire trend toward the industrialization and globalization of food production.

While the manifesto includes a critique of the dangerous direction in which we're headed, most important, it sets out a practical vision and ideas and programs to move us toward ensuring that food and agriculture become more socially and ecologically sustainable, more accessible, and for putting food quality, food safety, and public health above corporate profits.

We hope this manifesto will serve as a catalyst to unify and strengthen the movement toward sustainable agriculture, food sovereignty, biodiversity, and agricultural diversity, and that it will thereby help to alleviate hunger and poverty globally. We urge people and communities to translate it and use it, as appropriate to their needs, and to disseminate the principles and ideas it contains in as many ways as possible.

PART ONE
FAILURE OF THE INDUSTRIALIZED AGRICULTURE MODEL

The growing push toward industrialization and globalization of the world's agriculture and food supply imperils the future of humanity and the natural world. Successful forms of community-based local agriculture have fed much of the world for millennia, while conserving ecological integrity, and continue to do so in many parts of the planet. But these practices are being rapidly replaced by corporate-controlled, technology-based, monocultural, export-oriented systems. These systems of absentee ownership are negatively impacting public health, food quality and nourishment, traditional livelihoods (both agricultural and artisanal), and indigenous and local cultures, while accelerating indebtedness among millions of farmers and their separation from lands that have traditionally fed communities and families. This transition is increasing hunger, landlessness, homelessness, despair, and suicides among farmers. At the same time, it is degrading the planet's life-support systems and increasing, planetwide, the alienation of peoples from nature and the historic, cultural, and natural connection between farmers and all other people to the sources of food and sustenance. Finally, it helps destroy the economic and cultural foundations of societies, undermines security and peace, and creates a context for social disintegration and violence.

☓

Technological interventions sold by global corporations as panaceas for solving global problems of "inefficiency in small-scale production," and supposedly world hunger, have had exactly the opposite effect. From the Green Revolution to the biotech revolution to the current push for food irradiation, technological intrusions into the historic and natural means of local production have increased the vulnerability of ecosystems. They have brought the pollution of air, water, and soil, as well as new and spreading genetic pollution from genetically modified organisms (GMOs). These technology- and corporate-based monocultural systems seriously exacerbate the crisis of global warming because of their heavy dependence upon fossil fuels in all stages from production through distribution. Climate change alone threatens to undermine the entire natural basis of ecologically benign agriculture, bringing the likelihood of catastrophe to the near future. Moreover, industrial agriculture systems have certainly not increased efficiency in production if one subtracts their ecological and social costs and the immense public subsidies they require. Nor do these systems reduce hunger—quite the opposite. They have, however, stimulated the growth and concentration of a small number of global agriculture giants who now control global production, to the detriment of local food growers, food supply, food quality, and the ability of communities and nations to achieve basic food self-reliance.

℮

Already, negative trends of the past half century have been accelerated by the recent rules of global trade and finance from global bureaucracies like the World Trade Organization (WTO), the World Bank, the International Monetary Fund (IMF), and the Codex Alimentarius. These institutions have codified policies designed to serve the interests of global agribusiness above all others, while actively undermining the rights of farmers and consumers, as well as the ability of nations to regulate trade across their own borders or to apply standards appropriate to their communities. Rules contained in the WTO's Trade-Related Aspects of Intellectual Property Rights (TRIPS) agreement, for example, have empowered global agricultural corporations to seize much of the world's seed supply, foods, and agricultural lands. The globalization of corporate-friendly patent regimes has also directly undermined indigenous and traditional *sui generis* rights of farmers, for example, to save seeds and protect indigenous varieties they have developed over millennia.

Other WTO rules encourage export dumping of cheap sub-sidized agricultural products from industrial nations, adding to the immense difficulties of small farmers in poor countries to maintain their economically viability. And by invariably emphasizing export-oriented monocultural production, the explosive growth in the long-distance trade in food products has greatly increased the fossil fuels used for transport further impacting climate change. The expanding search for fossil fuels and fossil fuel alternatives has also resulted in ecologically devastating infrastructure developments in indigenous and wilderness areas, with grave environmental consequences.

⌀

The entire conversion from local small-scale food production for local communities to large-scale export-oriented mono-cultural production has brought the melancholy decline of the traditions, cultures, and cooperative pleasures and con-vivialities associated for centuries with community-based production and markets, diminishing the experience of di-rect food growing and the long celebrated joys of sharing food grown by local hands from local lands.

Despite all of the above, there are many developments that inspire optimism. Thousands of new and alternative initiatives are flowering across the world to promote ecologi-cal agriculture; defense of the livelihoods of small farmers; roduction of healthy, safe, and culturally diverse foods; and localization of distribution, trade, and marketing. Another agriculture is not only possible, it is already happening.

For all these reasons and others, we declare our firm opposition to industrialized, globalized food production and our support for this positive shift to sustainable, productive, locally adapted small-scale alternatives consistent with the following principles.

PART TWO
PRINCIPLES TOWARD AN ECOLOGICALLY & SOCIALLY SUSTAINABLE AGRICULTURE AND FOOD SYSTEM

THE ULTIMATE GOAL

The ultimate solution to the social, economic, and ecological problems cited above is a transition to a more decentralized, democratic and cooperative, noncorporate, small-scale organic farming method, as practiced by traditional farming communities, agroecologists, and indigenous peoples for millennia. Such communities have practiced a sustainable agriculture based on principles of diversity, synergy, and recycling. All rules and policies at every level of governance should be aligned to encourage such solutions as well as changes in other sectors of society to emphasize sustainability.

FOOD IS A HUMAN RIGHT

All human beings on the planet have a fundamental human right to access and to produce sufficient food to sustain their lives and communities. All rules and policies should be aligned to recognize this basic right. Every government—local, regional, national, international—is obliged to guarantee this right. It may not be denied in the interests of international commercial or trade processes, or for any other reason. Where a locality is unable to fulfill its obligation by reason of natural catastrophe or other circumstances, all other nations are obliged to provide the necessary help, as requested.

DECENTRALIZED AGRICULTURE IS EFFICIENT & PRODUCTIVE

We reject the notion that the globalization of industrial technological agriculture and the homogenization of farms brings greater efficiencies than local diverse community farming or the traditional agriculture deeply embodied in local cultures. Neither can industrial agriculture reduce world hunger. Countless experiences and studies show the opposite to be the case: the industrial monoculture system drives farmers from their lands, brings abhorrent external costs to the environment and to farming communities, and is itself highly susceptible to pests and a myriad of other intrinsic problems. Also, by most standards of measurement, small-scale biodiverse farms have proven at least as productive as large industrial farms. All policies at every level of society should favor small farms and the principles of agroecology to increase food security and insure robust, vital rural economies.

PUTTING PEOPLE, NOT CORPORATIONS, ON THE LAND

As the loss of small holder farmlands to wealthier landlords and global corporations is a primary cause of hunger, landlessness, and poverty, we support all measures to help people remain on or return to their traditional lands. Where peoples and communities have been deprived of their traditional lands and abilities to grow their own foods or to live in a self-sustaining manner, we strongly support distributive land reform to put people back on the land and the empowerment of local communities to determine their own destinies.

FOOD SOVEREIGNTY

We support the fundamental principle of national, regional, and community food sovereignty. All local, national, and regional entities and communities have the inherent right and obligation to protect, sustain, and support all necessary conditions to encourage production of sufficient healthy food in a way that conserves the land, water, and ecological integrity of the place, respects and supports producers' livelihoods, and is accessible to all people. No international body or corporation has the right to alter this priority. Neither does any international body have the right to require that a nation accept imports against its will, for any reason.

APPLICATION OF THE PRECAUTIONARY PRINCIPLE

All human beings have the right to food that is safe and nutritious. No technological interventions in food production should be permitted until proven to meet local standards of safety, nutrition, health, and sustainability. The precautionary principle applies in all matters.

CERTAIN TECHNOLOGIES DIMINISH FOOD SAFETY

Some technologies—such as genetic engineering, synthetic pesticides and fertilizers, and food irradiation—are not consistent with food or environmental safety. They bring unacceptable threats to public health, irreversible environmental impacts, and violate the inherent rights of farmers to protect their local plots from pollutants. As such, their use is incompatible with the viability of sustainable agriculture. No international body has the right to make rules that require that any nation accepts any foods or other agricultural imports across its borders that have been produced in this manner or that the nation considers detrimental to public health, the environment, local agriculture, cultural traditions, or in any other way.

IT IS IMPERATIVE TO PROTECT BIODIVERSITY & ECO-SYSTEM HEALTH

All healthy food and agricultural systems are dependent upon the natural world, with all its biodiversity intact. This protection of this biodiversity must be a priority for all governments and communities, and all policies should be aligned with this purpose, even where they imply changes in land tenure or farm size. No commercial or trade considerations or any other values may supersede this one. The principles of reducing "food miles" (distance food travels from source to plate), emphasizing local and regional production and consumption of foods, and reducing industrial high-input technological interventions are all derivative of the larger goal of environmental health and the vitality of natural systems.

THE RIGHT OF CULTURAL & INDIGENOUS IDENTITY

Agriculture and traditional systems of food production are an integral aspect of cultural identity, particularly among indigenous cultures; in fact, agro-biodiversity largely depends upon cultural diversity. All human communities have the right to preserve and further develop and enrich their diverse cultural identities, as historically practiced and expressed and passed on through generations. No international or national body has the right to alter these practices and values or seek to change them.

HUMANE TREATMENT OF ANIMALS

Industrialized "factory farming" and similar systems for beef, pork, chicken, and other animal production are notorious for inhumane conditions, as well as for their tragic ecological and public health consequences. Large-scale production increases the severity of the problems and brings with it the use of irradiation and antibiotic technologies to try and stem inherent problems of disease. All such practices must be banned, and all global and domestic rules that stimulate this manner of production must be actively opposed at every level of society.

THE RIGHT TO CONTROL & ENJOY LOCAL INHERITED KNOWLEDGE

All communities, indigenous peoples, and national entities have the inherent right and obligation to conserve their biological diversity and inherited local knowledge about food and food production, and to enjoy the benefits of this diversity and knowledge without outside interventions. This knowledge is key to preserving sustainable agriculture. All peoples also have the right to set their own goals for research and development, using local standards. No global-trade or intellectual-property-rights rule should require that local communities conform to any standards on these matters beyond their own. No global-trade rule or corporation should be allowed to undermine local farmers' or communities' rights to indigenous seeds or collective cumulative innovation and knowledge, or to promote "biopiracy," the robbing of local knowledge and genetic diversity for commercial purposes. Farmers' rights to save, improve, sell, and exchange seed is inalienable.

THE RELATIONSHIP BETWEEN FARMERS & THE ENVIRONMENT IS PRIMARY

We recognize, support, and celebrate the role of small-scale traditional and indigenous farmers as the primary sources of knowledge and wisdom concerning the appropriate relationship between human beings, the land, and long-term sustenance. Their direct experience of the nuances of interaction among plants, soil, climate, and other conditions and their crucial relationship with their communities must be protected, supported, and, where necessary, recovered. This historic role should no longer be threatened or interrupted by large-scale corporate systems run by absentee landlords operating on models that ignore local conditions and replace them with unworkable "one-size-fits-all" formulas.

THE RIGHT TO KNOW & TO CHOOSE

All individuals, communities, and national entities have an inherent right to all relevant information about the foods they consume, the processes used to produce them, and where the foods come from. This recognizes the sovereign right of people to make informed choices about the safety and health risks they are willing to take, both in terms of their own welfare and the environment's. This right notably applies to foods subjected to such technical interventions as pesticides, other chemicals, biotechnology, and food irradiation. No governmental entity, including international bodies, has the right to withhold information or to deny mandatory labeling and other disclosure of all risks, including those of malnutrition.

TRADE MUST BE VOLUNTARY, FAIR & SUSTAINABLE

We support the many diverse new trade initiatives within and among communities that are non-coerced, fair, sustainable, and mutually beneficial to producers and consumers, and under which communities voluntarily exchange goods and services of their own free accord and based on their own standards. No international body has the right to require that any nation or community must allow investment or trade across its borders, or the right to undermine local priorities. Every trade opportunity should be evaluated solely on its individual merits by each affected party.

NO PATENTS OR MONOPOLIES ON LIFE

We oppose the commercial patenting and monopolization of life forms. All international or national rules that permit such practices are violations of the inherent dignity of all life, the principles of biodiversity, and the legitimate inheritance of indigenous peoples and farmers worldwide. This applies to all plant life, animal life, and human life.

NO BIAS TOWARD GLOBAL CORPORATIONS

The inherent bias of international rule-making bodies such as the WTO and Codex Alimentarius toward large-scale, export-oriented monocultural production in agriculture, as in all other production, is a direct cause of social dislocation, environmental devastation, and the undemocratic concentration of global corporate power to the detriment of communities everywhere. All such rules should be immediately nullified and reversed to favor sustainable systems, local production, and local control over distribution. If these bodies refuse to make such changes, they should be abandoned as destructive to sustainable systems. Also, international bodies (such as the United Nations) should be encouraged to create new regulatory systems that act as effective international "antitrust" or anti-corporate-concentration institutions, in an effort to minimize corporate domination and its harmful effects.

FAVOR SUBSIDIARITY: A BIAS TOWARD THE LOCAL

Tariffs, import quotas, and other means by which nations attempt to further their own self-reliance, many of which have been made illegal or undermined by global bureaucracies, should be reinstituted to help reestablish local production, local self-reliance, and long-term food security. The principle of subsidiarity must apply. All rules and benefits should promote local production by local farmers, using local resources for local consumption. Trade will continue to exist but should consist mainly of essential commodities and other foods with unique appeal that cannot be grown locally. Long-distance trade must always be an available option, but not the *raison d'être* of the system. One imperative goal is a major reduction in overall long-distance trade, specifically in the distance between food producers and consumers (food miles).

SAFETY STANDARDS: FLOORS, NOT CEILINGS

All laws and rules concerning food agreed upon in bilateral and multilateral agreements among nations must reverse existing WTO priorities by creating a floor for safety standards, rather than a ceiling. No international body should make rules that require any nation or community to lower its own standards for trade or for any other reason. Such standards may include export and import controls, labeling, and certification. Any country or community with standards higher than those agreed upon by international bodies should receive favorable trade status. Poorer countries that cannot afford to implement such standards should receive financial aid to help them do so.

PROTECTION FROM DUMPING

The right to regulate imports to prevent dumping, to protect the livelihoods of domestic farmers, and to ensure a fair return for farmers' labor and contribution to food security is a fundamental element of just, fair trade rules. This reverses prior WTO rules that effectively permit and encourage dumping by large nations.

COMPATIBLE CHANGES ARE ALSO NEEDED

We recognize that the kinds of reform suggested above may be more rapidly achieved over time as part of a larger set of changes in prevailing worldview and systemic practices, so that ecologically and socially sustainable systems can take priority over corporate interests. Compatible changes may also be required in other operating systems of society, from global to regional, from corporate to community. Energy, transport, and manufacturing systems, for example, must be examined and reformed at the same time as farming recovers its small-scale, locally viable form. And all of these changes must come within the context of the principles of subsidiarity, which brings political power back from the global toward appropriate local and regional governance.

ADOPTION OF THESE PRINCIPLES

We urge all communities, municipalities, counties, provinces, states, nations, and international organizations to adopt these principles and to work in concert to realize them.

The following section gives examples of positive activities already underway that apply some of these principles, as well as specific proposals for new rules of trade governance consistent with them.

PART THREE
LIVING ALTERNATIVES TO INDUSTRIAL AGRICULTURE

On every continent, communities are awakening to the devastating effects of the corporate-driven food and farming systems that have turned agriculture into an extractive industry and food into a major health hazard. Movements are emerging, many with parallels and linkages across international borders, that are re-knitting the historic relationships among food, farming, and community values. These movements are restoring food and food production to their proper places in culture and nature after a devastating estrangement that stands as an aberration in the human experience.

Here we only have sufficient space to hint at the breakthroughs these movements have made in the past several decades. The fact that few of these changes could have been predicted should give pause to anyone who now argues that industrial agriculture is the inevitable way forward. Change—very rapid change—is possible. Indeed, it is underway. The following are a few of the areas where circumstances are rapidly changing:

DEMOCRATIZING ACCESS TO LAND

While it has long been recognized that access to land by the world's rural poor is a key to ending hunger and poverty, many believed reform to be politically impossible. This was true in Brasil, where less than 2 percent of rural landholders held half the farmland (most of it left idle) and where even small gatherings were outlawed and efforts for change were met with violence. Yet today this country leads the way toward democratizing access to land. During the last 20 years, the Landless Workers' Movement, called by its Portuguese acronym MST, has settled a quarter million formerly landless families on 15 million acres of land in almost every state of Brasil. Taking advantage of a clause in the new constitution mandating the government to redistribute unused land, the MST has used disciplined civil disobedience to ensure this mandate's fulfillment.

The MST's almost 3,000 new communities are creating thousands of new businesses and schools. Land reform benefits are measured in an annual income for new MST settlers of almost four times the minimum wage, while still-landless workers now receive on average only 70 percent of the minimum. Infant mortality among land reform families has fallen to only half the national average. Estimates of the cost of creating a job in the commercial sector of Brazil range from two to 20 times more than the cost of establishing an unemployed family on the land through land reform. Democratizing access to land is working.

DEMOCRATIZING ACCESS TO CREDIT

Bankers long held that poor people were unacceptable credit risks. But that barrier is falling. In Bangladesh two decades ago, the Grameen Bank created a rural credit system based not on property collateral but on small-group mutual responsibility. Grameen's micro-credit loan program, made to 2.5 million poor villagers, mostly women, has been adopted in 58 countries. Such loans boast a repayment rate far superior to those made by traditional banks. Democratizing access to investment resources is proving viable.

RE-LINKING CITY AND COUNTRY, CONSUMER AND GROWER

On every continent, practical steps are underway to make local production for local consumption viable. "Buy local" campaigns are appealing to consumers in Europe, the US, and elsewhere. One innovation is the Community Supported Agriculture (CSA) movement in which farmers and consumers link and share risks. Consumers buy a "share" from a farm at the beginning of the season, and receive regularly delivered boxes of vegetables throughout the season. CSAs emerged in the mid-60s in Germany, Switzerland, and Japan. Seventeen years ago, no CSAs existed in the US; today, there are more than 3,000 serving tens of thousands of families. The US example has helped inspire a CSA movement in the United Kingdom, which has won local government support. Similar movements have simultaneously developed elsewhere.

Another burgeoning initiative is urban and rural farmers' markets, which have grown by 79 percent in the last eight years in the US alone. These have enabled local farmers to sell directly to the public without expensive intermediaries. The cultivation of small garden plots—from family kitchen gardens in Kenya to school gardens in California in which children grow their own meals—is also spreading.

CREATING A RIGHT TO GOOD FOOD

Although 22 countries have enshrined the right to food in their constitutions, Belo Horizonte, Brasil's fourth largest city, is doing more. In 1993 its government declared food no longer merely a commodity but a fundamental right of citizenship. This shift did not trigger massive food handouts but ignited dozens of innovations that have begun to end hunger: patches of city-owned land are now available at low rent to local farmers as long as they keep prices within the reach of the poor; the city redirects the 13 cents provided by the federal government for each school child's lunch away from corporate-made processed foods and toward buying local organic food, resulting in enhanced nutrition. To enable the market to function more fairly, the city teams up with university researchers who, each week, post the sellers offering the lowest prices for 45 basic food commodities at bus stops and broadcast them over radio. These are only a few of the initiatives, all of which consume only 1 percent of the municipal budget. Officials from other Brasilian cities have come to Belo to learn from its successes.

ORGANIC & ECOLOGICAL FARMING IS SPREADING

Organic farming and grazing is spreading rapidly, now covering 23 million certified-organic hectares worldwide, with Australia, Argentina, and Italy in the lead. Defenders of the failing industrial, chemical approach to farming argue that organic farming can't work, but millions of sustainable farm practitioners are proving the naysayers wrong. Recent research examined more than 200 sustainable farming projects in 52 countries, encompassing approximately 70 million acres and 9 million rural farmers. This university-sponsored survey found that sustainable practices can "lead to substantial increases" in production. Some root crop farmers realized gains of as great as 150 percent using more-sustainable methods. Of course, with the much lower input costs of organic production, organic farmers often reap higher profits, even in rare cases where "yield" is slightly lower.

(In general, organic farming yields have proved higher in most cases when measured "per acre." Industrial systems, misleadingly tout yields "per worker," but in industrial systems, most workers are actually sacrificed to intensive machine and chemical production. The distortion in the measurement of the efficiency of industrial production is also magnified by the failure to account for "external" subsidized costs of environmental damage to land, soil, and public health.)

Increasingly, governments are providing direct support to organic farmers and to those converting, in order to meet growing consumer demand as well as for environmental and other benefits. In 1987 Denmark became the first country to introduce such national support; soon after that Germany began supporting conversion to organic farming. By 1996 all EU member states, with the exception of Luxembourg, had introduced policies to support organic farming. The Region of Tuscany, in Italy, has stood firmly against transgenic seeds and taken the lead in policies fostering small farms, ecological farming and local consumption. In Austria and Switzerland, 10 percent of farm production is organic, and in Sweden, 15 percent is organic. One Swiss canton has reached 50 percent organic production, and Germany's minister of agriculture has set a goal of 20 percent by 2010.

PROTECTING BIODIVERSITY

Internationally, the Convention on Biological Diversity now has 187 parties and 168 signatories. The Cartagena Protocol on Biosafety has been signed by 103 states. While multinational corporations have spread monocultures of small numbers of commercial, and now transgenic, seeds, a worldwide citizens' movement, working with responsive governments, is showing ways to protect seed diversity.

Citizen-education campaigns led by Greenpeace and others, for example, have contained GMOs to basically five countries: the US, Canada, Argentina, Brasil, and China. The Slow Food Movement, now with 80,000 members in 45 countries, is successfully reviving threatened seed varieties and generating renewed appreciation of local and regional food specialties. Spelt wheat, to pick just one example, the oldest known cereal, which was first cultivated in Italy in the Bronze Age but has been displaced in modern times by more commercial grains, is gaining consumers there. At the same time, indigenous peoples' movements are growing in the Global South to protect biodiversity, resist transgenic seeds, and oppose the patenting of life forms. Nayakrishi in Bangladesh, a movement of 50,000 farmers, is reviving traditional crops—saving, storing, and sharing seeds they carefully breed as the basis of household food security. In India, Navdanya, a project of the Research Foundation for Science, Technology and Ecology, has helped 100,000 farmers return to traditional, organic farming methods in villages now dubbed "freedom zones." The foundation and its network have successfully fought transgenic seeds and the patenting of indigenous knowledge. In large measure because of its efforts, Indian government officials recently refused to allow Bt cotton to be sold in the Punjab and other northern states after southern Indian farmers were hurt by its adoption.

ENSURING FAIR PRICES FOR PRODUCERS

A burgeoning worldwide fair trade movement is showing that the dominant system is not "free trade" and that a fair system is possible.

The fair trade movement began in Europe in the 1980s and has taken hold in 47 countries since. The system covers 12 products—most significantly coffee, on which 20 million households worldwide depend. Coffee bearing the "Fair Trade Certified" label, for example, has met a number of criteria concerning its production: the farmers who grew it received a floor price (now $1.26) no matter what the perturbations of the world market. The coffee is produced by democratically organized small farmers with full knowledge of market prices. In four years US demand for fair trade coffee has quadrupled to 10 million pounds. Worldwide fair trade, even in its short life, has kept an additional $18 million in the hands of producer families. The importance of fair trade cannot be overstated in a world economy where, in just one decade, the share of what is spent on coffee that returns to the producing countries has fallen from one-third to one-thirteenth.

Farmers are also successfully using producer cooperatives to reap a fairer return. Dairy cooperatives in Italy offer extensive varieties of dairy products. Today in India 75,000 dairy cooperative societies dot the country, with a membership of 10 million. The three biggest dairy businesses in the country are cooperatives, among them the Kaira District Cooperative Milk Producers' Union, born in 1946 in response to monopoly control over distribution and unfair return to producers. Similarly, in the US, Organic Valley, launched only 15 years ago with a handful of farmers, today has 519 member farmers and more than $125 million in sales. Last fall Organic Valley members in Wisconsin received almost twice the standard market price for their milk.

MAKING CORPORATIONS ACCOUNTABLE TO DEMOCRACY

Throughout the world, citizens are recognizing that huge global corporations with resources greater than most governments are essentially functioning as unelected public bodies. They must be brought within the controls of democratic governance, and there are significant movements to do so. For example, the majority of the world's governments have rejected the commercialization of genetically modified seeds. Even within the corporate-dominated US, nine states and two Pennsylvania townships now ban non-family-owned corporations from owning farms or engaging in farming. Additionally, a movement is beginning in the US that challenges the notion of "corporate personhood," which gives corporations constitutional rights overriding the rights of people and communities. Triggered by the ruinous effects of large hog-confinement operations, two municipalities in Pennsylvania now have ordinances denying corporations the constitutional protections of persons.

Some school districts in the US are rejecting the intrusion of corporate processed foods, which are tied to that country's epidemic of childhood obesity and diabetes. In a similar vein, localities in various parts of the world are rejecting the commodification of water.

AGRICULTURE BEYOND MARKET FUNDAMENTALISM

Such diverse but interrelated developments as indicated above point beyond "market fundamentalism," to the notion that all aspects of life should no longer be subordinated to global market considerations and the welfare of world-spanning corporations. In its place, these developments suggest a more open-ended democratic path. They point not to a new dogma, but to what many are calling "living democracy," suggesting that the well-being of all life must be taken into account. Living democracy, attuned to peculiarities of place and culture, flows from the essential engagement of citizens seeking solutions together and evolving with lessons learned.

PART FOUR
TRADE RULES TO ACHIEVE THE AIMS OF THE
INTERNATIONAL COMMISSION ON THE FUTURE OF FOOD

This section provides specific principles and suggestions for changes in the rules of the WTO so that they are consistent with the goals of the commission.

Current WTO trade rules have forced the continuous lowering of tariffs and other barriers that formerly protected the domestic economies of member nations. These more open borders have resulted in social and economic conditions that are detrimental to the majority, but to the benefit of large corporations. To achieve the aims of the commission, we advocate that these WTO rules must be replaced by new trade rules, to achieve the following goals:

PERMIT TARIFFS & IMPORT QUOTAS THAT FAVOR SUBSIDARITY

Most international trade rules now favor export production and the global corporations that dominate it. New rules must be instituted to permit the use of trade tariffs and import quotas to regulate imports of food that can be produced locally. They must emphasize support for local production, local self-reliance, and real food security. This means applying the principle of subsidiarity: all rules and benefits should promote local production by local farmers, using local resources for local consumption, thus shortening the distance between production and consumption. This is not to suggest that there should be no trade at all in food products. But trade should be confined to those commodities that cannot be supplied at the local level, rather than being the primary driver of production and distribution.

REVERSE THE PRESENT RULES ON INTELLECTUAL PROPERTY & PATENTING

The WTO has been attempting to impose the US model of intellectual property rights protection on all countries of the world. This model strongly favors the rights of global corporations to claim patents on medicinal plants, agricultural seeds, and other aspects of biodiversity, even in cases where the biological material has been under cultivation and development by indigenous people or community farmers for millennia. Most of these communities have traditionally viewed such plants and seeds as part of their commons, not subject to ownership or fee structures imposed by outside corporations.

These WTO rules on intellectual property should be abandoned. New rules should be adopted that favor the needs of local and domestic communities and the protection of innovation and knowledge developed over the centuries. New rules are also needed to deal with public health crises.

LOCALIZE FOOD REGULATIONS AND STANDARDS

With the false justification that they ensure food safety, many international rules, such as the WTO's Agreement on the Application of Sanitary and Phytosanitary Standards (SPS) and the Codex Alimentarius, have enforced a kind of industrial processing of foods that works directly against local and artisanal food producers, while favoring the global food giants. Among other things, the rules require irradiation, pasteurization, and standardized shrink-wrapping of certain products.

Such rules increase enormously the costs for small producers and also negatively affect taste and quality. In fact, the greatest threats to food safety and public health do not come from small food producers, but from large industrial farms and distributors. Their practices have accelerated the incidences of salmonella, E. coli, and other bacteria in foods, as well as mad cow and foot and mouth disease. Such homogenized, industrialized global standards have the primary goal of benefiting global corporate producers. We favor rules and food-production standards that are localized, with every nation permitted to set high standards for food safety.

ALLOW FARMER MARKETING-SUPPLY MANAGEMENT BOARDS

Currently disallowed by the WTO and North American Free Trade Agreement (NAFTA), these price and supply regulations let farmers negotiate collective prices with domestic and foreign buyers to help ensure that they receive a fair price for their commodities. Less than two years after NAFTA went into effect, Mexican domestic corn prices fell by 48 percent as a flood of cheap US corn exports entered the country. NAFTA dismantled the government price regulation agencies that could have maintained stable supply and prices for Mexico's domestic corn growers. Without these, thousands of farmers have been forced to sell their lands. Trade rules must allow the reinstatement of such practices.

ELIMINATING DIRECT EXPORT SUBSIDIES & PAYMENTS FOR CORPORATIONS

Although the WTO has eliminated direct payment programs for most small farmers, it continues to allow export subsidies to agribusiness. For example, the US Overseas Private Investment Corporation, funded by US taxpayers, provides vital insurance to US companies investing overseas. Even loans from the IMF to southern countries have been channeled into export subsidies for US agribusiness. Such subsidies help multinational corporations dominate smaller local businesses both domestically and abroad. All export subsidy policies should be eliminated. But programs that permit and encourage low-interest loans to small farmers and the creation of domestic seed banks and emergency food supply systems should be allowed.

RECOGNIZE & ELIMINATE THE ADVERSE EFFECTS OF WTO MARKET ACCESS RULES

Heavily subsidized northern exports to poor countries have destroyed rural communities and self-sufficient livelihoods throughout the South. Many people now working, for example, for poverty wages at Nike and other global corporate subcontractors are refugees from previously self-sufficient farming regions.

This entire model of export-oriented production is destructive to basic self-sufficient traditional farming. The dominant theory that exports from the South to North can be a major route for development ignores the inevitability of adverse competition between poor exporting countries for these rich markets and the hijacking of national priorities in the interest of cheaper exports. Also damaging to poor countries are the adverse working and environmental conditions demanded by the mobile corporations that dominate the global food export trade. To reverse this trend, countries must adopt new international trade rules that allow them to reintroduce constraints and controls on their imports and exports.

PROMOTE REDISTRIBUTIVE LAND REFORM

For the above changes in trade rules to really benefit the majority in a region, the redistribution of land to landless and land-poor rural families must be made a priority. This has been shown to be an effective way to improve rural welfare at different times in Japan, South Korea, Taiwan, and China. Research also shows that small farmers are more productive and more efficient and contribute more to broad-based regional development than do larger corporate farmers. Given secure tenure, small farmers can also be much better stewards of natural resources, protecting long-term productivity of their soils and conserving functional biodiversity.

Truly redistributive land reform has worked where it has been fully supported by government policies. These include debt-free government grants of land, full rights of title and use of land for women, the reallocation of only good-quality land, and easy access to predominantly local markets. The power of rural elites must be broken and reforms must be extended to the majority of the rural poor, so they have sufficient strength in numbers to be politically effective. There must be a highly supportive policy framework, reasonable credit terms and good infrastructure for sound local environment technologies.

CONCLUSION
SUMMARY OF TRADE-RULE CHANGES TOWARD ACHIEVING A SUSTAINABLE AND MORE EQUITABLE WORLD

The goal of these proposed global trade rules is to promote a more sustainable and equitable economic system by strengthening democratic control of trade and stimulating food and agricultural systems, industries, and services that benefit local communities, and re-diversifying local and national economies.

Protective barriers should be introduced to enable countries to attain self-sufficiency in food, where feasible, with long-distance trade primarily focused on providing food that cannot be grown in the country or region.

Quantitative restrictions that limit or impose controls on exports or imports through quotas or bans should be permissible. For those products that are imported, preferential access should be given to food, goods, and services going to and coming from states that, in the processes of production, provision, and trade respect human rights, treat workers fairly, and protect the environment.

Trade controls that increase local employment with decent wages, enhance environmental protection, ensure adequate competition and consumer protection, and otherwise improve the quality of life should be encouraged. States are urged to give favorable treatment to domestic food, products, and services that best further these goals.

States should make distinctions among the food and other products that they choose to import on the basis of the way they have been produced in order to further the aims of sustainable development.

Controls on trade should contribute to a wide range of purposes that further sustainable development, for example, sanctions against human rights violations; tariffs for the maintenance of environmental, food, health, and animal-welfare standards; enforcement of treaties on environment and labor rights.

All international laws and regulations that concern food and food safety and environmental and social standards should create a floor for governing the conditions for trade between parties. Any country with higher levels should receive favorable terms of trade. Poorer countries for whom such standards are too expensive should receive financial support to help them implement them. Once such a country has set a date for such improvements, it should receive the same favorable trade terms.

The "precautionary principle" is a justifiable basis upon which to establish regulatory controls affecting trade when the risks warrant action, even in the face of scientific uncertainty about the extent and nature of potential impacts.

Global patenting rights should not override the rights of indigenous communities to genetic and biological resources that are held in common. For food and other products, fees should be able to be charged to cover the cost of development, plus a reasonable level of profit, but such patenting rights must have a limited time frame and fully reimburse the parties whose knowledge contributed to the patented entity.

No individual investor may invoke international enforcement mechanisms against investment regulations of the nation states. The implementation of domestic investment regulations shall not be constrained by trade rules, provided that the investment regulations improve domestic social and environmental regulations and further such advances in trade relations.

Manifesto on the Future of Seed

IN 2003 THE INTERNATIONAL Commission on the Future of Food published and disseminated the Manifesto on the Future of Food. It laid out far-reaching concepts and practical steps toward ensuring that food and agriculture become more socially and ecologically sustainable and aimed to support the movement working for a more equitable and caring world. Translated into a number of languages, it has been widely disseminated to individuals and organizations, as well as at various conferences and gatherings, including the World Trade Organization (WTO) Ministerial in Cancún, Mexico, in 2003 and has been adopted by different communities throughout the world. Out of its holistic vision and principles the dire state of seed, with all its ramifications, has emerged as an imperative that must be addressed.

With the continued support and active participation of the government of the Region of Tuscany, the International Commission on the Future of Food, through a global stakeholder consultation at Terra Madre 2006 in Turin, has prepared the present Manifesto on the Future of Seed.

We hope this manifesto can serve to further strengthen and accelerate the movement toward sustainable agriculture, food sovereignty, biodiversity, and agricultural diversity; help defend the rights of farmers to save, share, use, and improve seeds; and enhance our collective capacity to adapt to the hazards and uncertainties of environmental and economic change.

We urge people and communities to use it as appropriate to their needs and as a tool to unify and strengthen the call to counter the threat to seed and biodiversity imposed by industrial agriculture and multinational corporate interests.

PART ONE
DIVERSITY OF LIFE AND CULTURES UNDER THREAT

Seeds are a gift of nature, past generations, and diverse cultures. It is our inherent duty and responsibility to protect and to pass seeds on to future generations. Seeds are the first link in the food chain, the embodiment of biological and cultural diversity, and the repository of life's future evolution.

Since the onset of the Neolithic revolution some 10,000 years ago, farmers and communities have worked to improve agricultural yield, taste, and nutritional value. They have expanded and passed on knowledge about the health impacts and healing properties of plants as well as about their peculiar growing habits and their interaction with other plants, animals, soil, and water. Rare initial events of hybridization have resulted in larger-scale cultivation of certain crops in their centers of origin (such as wheat in Mesopotamia, rice in Indochina and India, and maize and potato in Central America), which have since spread around the globe.

Throughout this period the free exchange of seed among farmers has been the basis of maintaining biodiversity as well as food security. This exchange is based on cooperation and reciprocity, in which farmers generally exchange equal quantities of seed. And the exchange goes beyond the actual seed: it extends to the sharing and exchange of ideas and knowledge, of culture and heritage. It is an accumulation of tradition, of knowledge of how to work the seed gained by farmers actually watching the seed grow in each other's fields. The cultural and religious significance of the plant, its gastronomic value, its drought and disease resistance, its pest resistance, its ability to be saved, and other characteristics shape the community's knowledge of the seed and the plant it produces.

Today the diversity and future of seed is under threat. Of 80,000 edible plants used for food, only about 150 are being cultivated, and just 8 are traded globally. This implies the irreversible disappearance of seed and crop diversity.

The erosion of diversity has been propelled by industrial agricultures' drive for homogenization. The freedom of seed and the freedom of farmers are threatened by new property rights and new technologies that are transforming seed from a commons shared by farmers to a commodity monopolized by corporations.

Similarly, the rapid extinction of diverse crops and crop varieties and the development of nonrenewable seeds, such as property hybrids and sterile seeds based on the "terminator technology," threatens the very future of seed, and with it the future of farmers and food security.

✸ EROSION & EXTINCTION OF DIVERSITY

The acceleration of technological revolutions in all fields and the growing concentration of economic power in the hands of a small number of people and organizations have produced an increasing homogenization of production strategies and of human cultures in our world. As a result, the genetic variability of domesticated and wild plants and animals and the diversity of languages and cultures are being destroyed at an unprecedented level.

At the same time, industrial production strategies have unleashed unexpected long-term effects on the climate and on the whole network of life systems. This process of ecological destruction and genetic erosion has been accelerating over the past decades. As a consequence of this human activity, abrupt and profound eco-systematic planetary changes within the present century can be foreseen.

Furthermore, industrial production has not only made abrupt and profound change an impending certainty, but it is also destroying the very diversity that is the only proven strategy of living beings to cope with such change. While plants, animals, and microorganisms make use of their genetic variability, humans depend on their cultural variability and their inventive capacity to adapt to changes in the environment around them in order to survive.

These destructive industrial agricultural practices, as well as wars and expulsion, are reducing seed diversity more dramatically than ever before.* The biased usage of unexpected advances and successes in biology, particularly in genetic and molecular biology, has played a significant role in this decimation. Technologies such as chemical fertilizers and genetically engineered crops derived from now obsolete interpretations of biological concepts have been developed and advertised as the only way to overcome worldwide problems like famine and illness and are used as tools for economic and political control. The disappearance of local seeds has gone hand in hand with the disappearance of small farmers and local food cultures. And with them, local knowledge about the use of cultivated and wild plant varieties in their different ecological and cultural habitats has likewise been lost. With the extinction and reduction of languages and cultures the indigenous names for and distinctions among thousands of plants have been lost, as have the experiences and traditions of their cultivation.

---

* Crop genetic resources are disappearing at the rate of 1 to 2 percent a year (United Nations Food and Agriculture Organization [FAO] Development Education Exchange Papers, September 1993). About 75 percent of the diversity of agricultural crops is estimated to have been lost since the beginning of the last century.

Civilizations rose with new agricultural technologies. The ability to produce more food than needed by those working in the fields was key to the development of progressively more sophisticated divisions of labor. Traditionally, in most rural communities, the selection, preservation, and maintenance, the wise development and passing on of seed stock has been—and is still today—the domain of women. Preserving seed for the next season has been a fundamental rule of survival in human history.

Systems of rights and responsibilities must be evolved that both recognize the collective rights of local communities and the seed sovereignty of farmers, as well as the mutual interdependence among diverse cultures and countries.

THE BIAS OF INDUSTRIAL AGRICULTURE & SEED BREEDING

Industrial agriculture has severely eroded the biological diversity of seeds, crops, and breeds of livestock. The spread of modern commercial agriculture and the replacement of local varieties has been identified as the chief contemporary cause of the loss of genetic diversity[*] and the most important cause of genetic erosion.[†]

Industrial agriculture, for which the lion's share of commercially traded seeds is produced, relies on a production process that conflicts with basic rules of seed production and

---

[*] Stated in the Leipzig Global Plan of Action on Plant Genetic Resources for Food and Agriculture, 1995, based on 158 country reports and 12 regional and subregional papers.

[†] FAO Leipzig Conference on Plant Genetic Resources, 1996.

reproduction. The goal of ever-increasing yields of individual commodities comes at reduction of overall output and erosion of biodiversity. It is driven by short-term managerial concerns and profit margins and by its very nature sacrifices consideration of the public good, such as the long-term sustainability of soil, ecosystems, and farming communities.

This market-driven approach is often reflected at the government level. Rather than acting in the interest of the public good, governments further distort market prices by granting subsidies to domestic companies, giving them a competitive advantage and thereby reducing prices. These artificially low prices are pushing both biodiversity and small farmers to extinction.

It is obvious and generally accepted that such industrial agriculture and commodity market policies further deplete our already limited natural resources, increase energy and toxic inputs at the expense of labor, and lead to rural despair and hunger in the world. This despite the fact that more agricultural products are produced than are needed to feed all 6.5 billion citizens of this planet—and, if wisely distributed, enough is already produced to feed the additional 2.5 billion people expected to swell the global population in the next 40 to 50 years. The inadequacy of the current model of food production is evident from the fact that while more than a billion people are hungry and suffer from malnutrition due to being underfed another 2 billion suffer malnutrition due to being overfed with unhealthy food. For the first time, the number of children suffering from obesity is about some GM crops showed that it is even hard to control such traits within the commercial product chain. Ordinary seeds are frequently contaminated with GM traits in areas where genetically modified organisms (GMOs) are planted. This poses a massive and immediate threat to farmers wishing to continue producing GMO-free products in response to the growing rejection of GM foods by consumers worldwide. So far, only two GM traits have gained significant market share, one conferring resistance to a broad spectrum herbicide Roundup (RR) and another making plants poisonous to insects by means of a soil microbe Bacillus thuringiensis (Bt). Within a few years plants with these GM traits—soybeans, maize, oilseed rape, and cotton—have grown to cover an area of about 90 million hectares annually, concentrated in five "GM countries" (the US, Canada, Argentina, Brasil, and China plant more than 90 percent of the world's GM crops). The impact of these GMOs on seed diversity as well as on the overall biodiversity in those areas is devastating. A single multinational company, Monsanto, holds the patents for 90 percent of all commercial GM plant traits.

✂ CORPORATE TAKEOVER OF SEED: A THREAT TO SEED FREEDOM & THE RIGHTS OF FARMERS

Until very recently, seed has resisted basic principles of capitalist market laws, the most important barrier being the nature of the to outnumber those children suffering from hunger.

One characteristic of this "mechanistic utopia," which reduces living systems to machines whose output can be maximized and strives for "the best" of all crops and varieties, is the attempt to adapt environmental conditions to the production system rather than adapting production to different ecosystems and cultural traditions. Such attempts have a devastating effect on the environment, natural resources, and on the rural communities subjected to them. The "Green Revolution," which was probably the single most dramatic boost in caloric yields per acre in recent history, is the iconic example of what can go wrong with such apparently successful linear and productionistic improvements. Today it is apparent that the nutritional impact, especially on rural populations and the poor in those regions that were to benefit most from the Green Revolution, has in fact been largely negative.

GENETIC ENGINEERING

In the mid-1990s, the first genetically engineered seeds were commercialized. Genetic engineering technology transfers the DNA sequences for individual traits in ways that could not occur naturally. The risks involved in this technology for human health and the environment and especially the long-term effects on biodiversity are unpredictable. Once released into the environment, these genetically modified (GM) plants reproduce and outcross to wild relatives and it becomes impossible to recall them. Scandals over the illegal release of seed, which reproduces itself and multiplies. Thus, seed has long been both a means of production as well as the product itself.

Research and development for seed improvement has long been a public domain and government activity for the common good. However, private capital started to flow into seed production and took it over as a sector of economy, creating an artificial split between the two aspects of the seed's nature: its role as means of production and its role as product. This process gained pace after the invention of hybrid breeding of maize in the late 1920s. Today most maize seed cultivated are hybrids. The companies that sell it are able to withhold the distinct parent lines from farmers, and the grain that it produces is not suited for seed saving and replanting. The combination guarantees that farmers will have to buy more seed from the company each season. In the 1990s the extension of patent laws as the only intellectual property rights tool into the area of seed varieties started to create a growing market for private seed companies. Previously, intellectual property rights had had a much milder effect on the seed market. They were based on the concept of plant-variety rights, under which the farmer could use purchased seed for further sowing and breeding and could freely use the yield of seed for saving and replanting. The farmer was only restricted from commercially reselling the saved seeds.

INTELLECTUAL PROPERTY RIGHTS & SEED MONOPOLIES

The advent of genetic engineering led to the almost-worldwide introduction of industrial patents on life forms. These patents put exclusive and total private control over discoveries, now redefined as "inventions," into private hands. Under these patent laws, seeds are entirely subordinated to a system of intellectual property rights (IPRs), which by law—though not necessarily in reality—convert such seeds into nonrenewable production inputs that farmers must repurchase every year. In addition, over the past two decades, hybrid seed production has been extended to plants previously inaccessible to this technology. This technology has reached its ultimate development with the advent of "terminator seeds" (also called GURTs), which produce seed that is sterile or suicidal by nature or only reproduces upon the addition of certain external inputs. Meanwhile, seeds as well as individually isolated DNA sequences have become subject to industrial patenting. Plant-variety protection under the Union for the Protection of New Varieties of Plants (UPOV) system has also expanded to include fees for replanting of seeds and to incorporate industrial patent rights on GMOs. The WTO, under its Trade Related Intellectual Property Rights (TRIPS) agreement, obliges member states to introduce general IPR systems on plants. In addition, following the breakdown of the WTO talks in July 2006, industrialized countries have been intensifying the imposition of IPR laws on developing countries through bilateral trade agreements. These are further undermining the potential of the Convention on Biological Diversity (CBD) and the International Treaty on Plant Genetic Resources for Food and Agriculture, an international agreement to secure exchange of seed under the emerging global IPR regime.

The review of the TRIPS agreement, including Article 27.3(b) on plants, seed and biodiversity set for 1999 has been systematically ignored. Formal submissions have been made by many countries of the South to exclude life forms, including seeds, from patenting. This review of TRIPS, neglected for too long, must be undertaken as a matter of the highest priority.

PRIVATIZATION OF SEED

The artificial split of seed into production tool and product and its transformation into a pure commodity have been extended to most areas of industrial agriculture today. Though this process has not taken place without controversial discussions and fights, especially in rural areas of developing countries, an unprecedented global concentration of private seed companies is taking place. Small seed companies as well as entire national seed collections and institutions are being bought up for comparatively moderate prices by agro-chemical multinationals. For these companies, seeds are but one component of their sales packages of agricultural and chemical input and a further strategy to vertically integrate the global market of agricultural commodities, whether used as food or for nonfood purposes.

The transformation of a common resource into a commodity, of a self-regenerative resource into mere "input" under the control of the corporate sector, changes the nature of the seed and of agriculture itself. It robs peasants of their means of livelihood, and the seed becomes an instrument of poverty and underdevelopment, one that has displaced huge numbers of farmers.

Public funding for seed development and conservation has been steadily dwindling. It has reached such low levels that even major seed collections are under threat and increasingly depend upon so-called public-private partnerships. Such partnerships open the way for private seed companies to further expand their IPR-based control over the global seed stock. While public seed collections are obliged to provide free samples of their holdings, private companies can choose not to participate in this system of free exchange and abuse it for their own interests. In addition, every new step of corporate concentration of seed stocks brings about a reduction of seed varieties as well as a reduction of the number of breeders and scientists maintaining these seed stocks. There is a clear relation between the increase in investment in the digitalization of seed information at the DNA and genomic level and a parallel decrease in investment in on-field research and the development and maintenance of holistic research and knowledge of seed and seed varieties in different ecosystems.

PART TWO
A NEW PARADIGM FOR SEED

A post-industrial concept of seed and food production must take into account the failures, limitations, and vulnerability of industrial agriculture and corporate monopolies. It must be based upon holistic, long-term considerations—ones that industrial agricultural systems producing for a global market, by their very nature, cannot address.

Seed diversity can be maintained only if the livelihoods of small farmers who save and use biodiversity are protected. Biodiversity-based farming systems generate more employment, produce more nutrition and better quality food, and provide higher incomes to farming families. The goal of agriculture must no longer be to produce huge quantities of nutritionally unbalanced food, but rather to produce nutritionally balanced food in a sustainable way, one that preserves natural resources and the communities' social and cultural systems, which allow for the appropriate distribution of food, and one that provides the possibility of a decent livelihood in rural areas.

The one-dimensional focus on yield has led to a serious decline in systems productivity, food quality, and nutrition. Quantity must give way to quality. Seed production by food communities is based on a holistic concept of food quality that considers taste, compatibilities with human physiological and cultural conditions, all aspects of nutrition, the degree of biodiversity, and the environmental impact of production, as well as the working conditions, processes of participation, and value of contribution of producers. This holistic concept should be the basis for reinforcing or creating quality seed and food systems.

Any future concept of agricultural production must anticipate and take into account the change in climatic conditions and urgently introduce stringent measures to reduce CO_2 and other greenhouse gas emissions.

The monoculture paradigm must give way to a flourishing biodiversity paradigm. In addition, we must address the rapidly expanding water crisis, which may be dramatically exacerbated by climate change. Drinking water is already scarce in many regions of the world, and we must make sustainable freshwater management a priority. A sustainable water-management plan must also stop the ongoing soil erosion to preserve the basis of agricultural production and must phase out the alarming input of toxic substances into vital ecosystems as well as the human food chain.

Reducing the waste of energy and natural resources due to irrational, counterproductive, unhealthy systems of processing, storage, transport, and consumption must become an integral part of future plans for sustainable food production and consumption policies.

Finally, plans for sustainable agricultural production must aim at reducing and ideally stopping the present unsustainable rate of urbanization and development of megacities. They are devastatingly destructive to the ecology and create high-risk hot spots prone to the destructive power of turbulent climate change.

International agreements such as the Food and Agriculture Organization (FAO) Plant Genetic Resources for Food and Agriculture, and the Convention on Biological Diversity—which recognize the need to conserve biodiversity and defend farmers' rights, as well as national and subnational laws that have upheld the rights of farmers to save, use, exchange, improve, and develop seeds—need to be upheld and strengthened and made effective instruments to counter the growing corporate monopoly over seeds.

It is at the local level that the new seed paradigm is taking shape. Communities are creating movements to save and share seeds and create alternatives to unsustainable agriculture based on monocultures and monopolistic "intellectual property rights."

PART THREE
THE LAW OF SEED

Diversity, freedom, and ensuring the potential of future evolution of agriculture and humanity are core principles of the law of seed.

ℒ DIVERSITY

Diversity is our highest form of security. Diversification has been the most successful and widespread strategy of agricultural innovation and survival over the past 10,000 years. It increases the array of options and the chances of adapting successfully to changing environmental conditions and human needs. For these reasons and others, and in contrast to the present trend toward monocultures and genetic erosion, diversity must once again become the over-arching development strategy in the following ways:

DIVERSITY OF SEED

There is an urgent need to maintain seed diversity, to expand the number of plants used for human nutrition, and to expand the varieties cultivated within a given plant species. Reversing the trend toward monoculture is one of the most urgent tasks we face if we want to preserve our chances of adapting and surviving changing conditions in the years and millennia ahead.

DIVERSITY OF AGRICULTURAL SYSTEMS

Agricultural policies aimed at promoting and implementing global diversity of seed cultivars must support the development and spread of holistic agricultural systems, in which human, crop, animal, and microbial biodiversity are employed as indispensable tools to reduce external inputs, to increase productivity efficiency, and to achieve sustainability.

Two main categories have to be considered:

- Traditional low-external-input agricultural systems, in which crop biodiversity (poly-cultures) and seed mixtures (consociations) help to fulfill farmers' needs at different levels of production
- Ecological agricultural systems, in which seed diversity is used to maintain planted biodiversity (crop rotation) and associated biodiversity (soil, plants and fauna)

DIVERSITY OF PRODUCER-CONSUMER RELATIONSHIPS

Agricultural biodiversity is best conserved when farmers are able to earn a decent income. The consolidation of the production and distribution system of food depletes biodiversity. Food systems in which producers have direct contact with consumers enrich biodiversity. Diversity of producer-consumer relationship is key to food democracy and protection of biodiversity.

DIVERSITY OF CULTURES

Biodiversity and cultural diversity go hand in hand. Preserving, maintaining, and expanding the remaining agricultural traditions and cultures of production is an immediate and urgent task if we are to prevent the further erosion of biodiversity. Such work entails respect and appreciation of the different traditions and ways that humans perceive nature and food cultures.

DIVERSITY OF INNOVATION

Hundreds of thousands of communities and farmer cooperatives, millions of family and subsistence farms, and gardeners around the world form the basis not only for conservation and propagation of farmers' varieties but also of further development of seed. The addition of scientists and professional plant breeders to the art of participatory plant breeding would make an even more formidable force for innovation. Finding fair and equitable ways for these different groups to cooperate and integrating their diverse levels of knowledge and experience would give enormous impetus and strength to our ability to meet future challenges.

✑ FREEDOM OF SEED

Seeds are a gift of nature and diverse cultures, not a corporate invention. Passing on this ancient heritage from generation to generation is a human duty and responsibility. Seeds are common property, to be shared for the well-being of all and saved for the well-being of future generations and hence cannot be owned and patented. Seed saving and sharing is an ethical duty that should not be interfered with by national or international laws which try to make seed saving and seed sharing a crime.

The "law of the seed" must protect the freedom of seed and the freedom of farmers, based on the following principles:

FREEDOM OF FARMERS TO SAVE SEEDS

The first duty and right of farmers is to protect and rejuvenate biodiversity. The conservation of biodiversity requires the saving of seed. Laws of compulsory registration and policies for "seed replacement" undermine the freedom of farmers to save seeds. Intellectual property laws, patent laws, and breeders' rights laws violate the law of the seed by making it illegal to save seeds.

FREEDOM OF FARMERS TO BREED NEW VARIETIES

Farmers' rights derive from their intellectual contributions to the breeding of seeds and plant genetic resources. Though their breeding objectives and methods might differ from the objectives and methods of the seed industry, farmers are breeders. Farmers breed for diversity while the seed industry breeds for uniformity. The recognition of farmers' breeding strategies is necessary to stop the practice of using farmers' seeds as "raw material" with no compensation for the intellectual contribution of farming communities and to ensure the ability of farmers to further develop new varieties. Farmers have the right to freely develop new varieties of seeds.

FREEDOM FROM PRIVATISATION & BIOPIRACY

Farmers' innovation in plant breeding takes place collectively and cumulatively. Therefore farmers' rights arising from their role as conservers and breeders have to be vested in farming communities, not in individual farmers.

The recognition of farmers' collective rights is necessary to protect seeds and biodiversity as a commons. It is also necessary in order to stop the seed industry's practice of using farmers' varieties as "raw material" and then claiming patents and intellectual property rights on the basis of invention of the traits derived from them, a phenomena referred to as biopiracy. The global seed industry misuses the concept of the "common heritage of mankind" to freely appropriate farmers' varieties, convert them into proprietary commodities, and then sell them back to the same farming communities at high costs and with heavy royalties. Such privatization through patents and intellectual property law violates the rights of farming communities and leads to the indebtedness, impoverishment, and dispossession of small farmers.

Farmers' and food communities' access to seeds and plant genetic resources must not be restricted by private property claims and patent laws. Farmers should have access to their seeds in gene banks across the world. This freedom is the basis of farmers' seed sovereignty.

FREEDOM OF FARMERS TO EXCHANGE & TRADE SEEDS

Since seeds are a commons, freedom to exchange seeds among farming communities must be an inalienable part of the law of the seed. This also includes the right to sell and to share seeds on a nonexclusive basis. Any price paid for seeds should be calculated as a fraction of the value of the products they yield.

FREEDOM TO HAVE ACCESS TO "OPEN SOURCE" SEED

"Open Source" seeds are open pollinated varieties, which can be reproduced from year to year, and from generation to generation, and can be saved and replanted. The knowledge about the information embedded in seeds and germ plasm is, by its nature, not an invention but the result of cumulative collective discovery, a common effort upon which future discoveries may be based. This knowledge should be freely available and made accessible to all farmers. Seed systems that cannot be reproduced by farmers should not be developed. On the contrary, research and development should concentrate on seeds that can be freely reproduced. Public investment should go exclusively into seed systems that contain the full genetic information necessary for their reproduction. Farmers should have access to parent lines used for crossing and the creation of hybrids. Corporate control of hybrid parental lines leads to homogenization and monopoly ownership.

FREEDOM FROM GENETIC CONTAMINATION AND GMOS

Farmers' freedom includes freedom from genetic contamination and biopollution. The introduction of new varieties and plants must take into account the potential environmental risks as well as other potentially detrimental agricultural effects.

FREEDOM OF SEED TO REPRODUCE

Terminator technology to produce sterile and suicide seed violates the freedom of seed to reproduce. The production of seed that cannot reproduce is an assault to the fundamental nature of seed as the source of reproduction of life and to the fundamental freedom of farmers. The introduction of such traits is designed to create a monopoly on the seed and food of the world. It must be banned on a global level.

✿ SEEDS FOR THE FUTURE: BREEDING TOMORROW'S SEEDS

Seeds embody the past and the future. The evolution of seeds in the future depends on our conserving the widest diversity of seeds and crop varieties to deal with the multiple challenges of food and nutritional security, food quality, climate change, and sustainability.

The following are ways in which the conservation, use, and further development of seeds can be tailored to meet the challenges ahead:

COMMUNITY-BASED SEED CONSERVATION & DEVELOPMENT

The preservation and maintenance of seeds and the knowledge about them should be based on and rooted in those who make use of them. Ex-situ and in-situ conservation of germ plasm should be conducted to support essential on-farm seed maintenance. Strategies and technologies for the further development of seeds should be based on the wealth of experience and ingenuity of farmers and food communities in general and include their participation and active input. This process should include making modern technologies of selection, identification, and breeding available to farming communities.

EMBEDDING IN AGRICULTURAL ECOSYSTEMS

As a matter of principle, seed varieties should be developed that allow farmers to conserve soil, water, and biodiversity and intelligently adapt to local and regional environmental conditions, rather than require the adaptation of the environment to the needs of the seed. We should develop the seeds of tomorrow with the goal of embedding agricultural production in agro-ecosystems to protect soil, water, and biodiversity and increase resilience to environmental change.

REDUCE GREENHOUSE GAS EMISSIONS

To minimize the emission of the greenhouse gases that are leading to climatic chaos, seeds should require no more external input of energy (through synthetic chemical fertilizers, pesticides, and fuel) than absolutely necessary. The goal should be agricultural practices that are greenhouse-gas-emission neutral and that rely on renewable energy and on soil-biological resources.

PHASE OUT & ELIMINATE TOXIC INPUTS

To reduce the toxic contamination of our food chain and environment, we should shift from breeding seeds that respond to chemical inputs to seeds that are better adapted to the requirements of agroecological practices.

DIVERSITY WITHIN VARIETIES

As a means to reduce the risk of susceptibility to pests and adverse environmental conditions, and to enhance the natural diversity, future seed development should be based on the broadest possible genetic diversity. To this end, an urgent review of present commercial requirements for the homogeneity of seed varieties is called for.

BREEDING FOR FOOD QUALITY

The holistic quality of food, including its taste and nutritional value, should be the dominant concern for further enhancing, preserving, and developing seeds of the future.

WOMEN ARE THE PROTAGONISTS OF BIODIVERSITY

Globally, women represent the majority of the agricultural workforce and are the present and traditional custodians of seed security, diversity, and quality. Women are also the prime repositories and disseminators of knowledge about the quality and methods of processing food. As such, their central role in safeguarding biodiversity and in conserving, exchanging, and reproducing seeds in post-industrial agriculture must be supported and enhanced.

PART FOUR
LIVING ALTERNATIVES—SEEDS OF HOPE

Seeds are an expression of hope. They bring to mind a cornucopia of harvest. Large numbers of individuals, initiatives, and traditional food communities the world over have long been engaged in safeguarding seed. Despite the present alarming scenario of monocultures and corporate seed monopolies, many encouraging initiatives have sprung up to counter the threat posed by industrial agriculture. The principles on which this manifesto is based have evolved from the initiatives and actions of diverse groups and movements across the world. The following are some such examples.

- A mushrooming of seed banks to preserve ex-situ and cultivate in-situ seed and plant diversity is taking place within seed communities. Women have played a pivotal role in safeguarding the heritage of seed and are set to continue to do so in increasing numbers. Movements such as Seeds of Survival in Ethiopia and Navdanya in India have evolved new models of saving seeds and enhancing the food security and ecological security of farming communities.

- Seed-saving initiatives and seed-exchange platforms are taking on an increasingly important role. Large numbers of individuals are creating gardens with the express purpose of growing their own food. They have the potential to play an important role in seed saving and exchange. A number of communities committed to the protection of, and reversing the huge losses in, seed and breed varieties are rallying their forces. One such example is the Presidia biodiversity-protection project of the Slow Food Foundation for Biodiversity. Versions of this project have sprung up in all regions of the world.

- Targeted plant-breeding projects that adapt seeds to the needs of organic and ecological agriculture are fast increasing.

- Alliances and networks of civil society around seed are being built on the regional, national, and international levels. Examples include networks like ETC and GRAIN and political-pressure initiatives like Save Our Seeds, as well as farmer rights groups.

- Movements such as No Patents on Life in Europe and movements to create patent-free zones (Living Democracy/Jaiv Panchayat) and non-cooperation with patents on seeds (Bija Satyagraha) in India, and the seed-sovereignty movement of Native American tribes in North America are evolving from the ground up to defend the freedom of seed.

- Working in parallel with civil society activities are initiatives to adopt laws and that establish GMO-free zones on a large scale and protect diversity of seed. The region of Tuscany's Law on Seed is a good example of how local and regional governments can take responsible and concerted action to protect seed diversity.

- The fast-growing direct relationships between producers and consumers, such as Community Supported Agriculture (CSA) networks, are another vibrant step in the movement toward conserving and maintaining seed and plant varieties.

- International agreements such as the Treaty on Plant Genetic Resources for Food and Agriculture and its Article 9 on Farmers' Rights and the Convention on Biological Diversity have the potential to counter the aggressive control and suicide-oriented policies of large multinational corporations. This potential needs to be strengthened.

- Demands to review Act 27.3(b) of the TRIPS agreement of the WTO and stop the patents on life, patents on seeds and biopiracy of farmers' varieties and traditional knowledge continue to be made by third world governments.

The future evolution of humanity goes hand in hand with the future and free evolution of our seeds. What is embedded in and has been practiced by peasant cultures from time immemorial needs the utmost support from the public and private sector if our right to choose and to live healthy, safe, and culturally diverse lives is to prevail.

The future of seeds carries within it the future of humanity.

APPENDIX
INTERNATIONAL COMMISSION ON THE FUTURE OF FOOD AND AGRICULTURE

A JOINT INITIATIVE OF:

Claudio Martini, President of the Region of Tuscany, Italy, and **Vandana Shiva,** Executive Director, Research Foundation for Technology, Science and Ecology, Navdanya, India

AND

Jerry Mander, Editor, "Manifesto on the Future of Food," and President of the Board of the International Forum on Globalization

PARTICIPANTS AT COMMISSION MEETINGS:

Vandana Shiva, Chair

Miguel Altieri, Professor, Department of Environmental Science Policy and Management, University of California at Berkeley

Debi Barker, Co-Director and Chair of the Agricultural Committee of the International Forum on Globalization

Aleksander Baranoff,† President, ALL, National Association of Genetic Safety, Moscow

Wendell Berry, Conservationist, Farmer, Author, and Poet

Marcello Buiatti, Consultant on GMO issues to Tuscany, Professor, University of Florence

Jose Bové,[†] Campesina, France

Peter Einarsson, Swedish Ecological Farmers Association, International Federation of Organic Agriculture Movements, EU Group

Elena Gagliasso, Scientific Coordinator for the Lega Ambiente, Professor, University of Rome

Bernward Geier, Director, International Federation of Organic Agriculture Movements

Edward Goldsmith, Author, Founder and Editor of *The Ecologist*

Benny Haerlin, Foundation of Future Farming, Germany, Former International Coordinator of GMO campaign for Greenpeace

Colin Hines, Author of *Localization: A Global Manifesto*; Fellow, International Forum on Globalization

Vicki Hird, Policy Director, Sustain: The Alliance for Better Food and Farming

Andrew Kimbrell, President, International Center for Technology Assessment

Tim Lang, Professor of Food Policy, Institute of Health Science, City University, London

Frances Moore Lappé, Author, Founder, Small Planet Institute

Caroline Lucas, Member of the European Parliament, Green Party United Kingdom

Jerry Mander, President of the Board of the International Forum on Globalization

Helena Norberg-Hodge, International Society for Ecology and Culture

Carlo Petrini, Slow Food, Italy

Assétou Founé Samake,[†] Biologist, Geneticist, Professor, Faculty of Sciences, University of Mali

Kristen Corselius,[§] Institute for Agriculture & Trade Policy, USA

Raj Patel,[§] Food First, USA

Sandra Sumane, Sociologist at the University of Latvia, Riga

Percy Schmeiser, Farmer and GMO activist

Alice Waters,[†] Founder, Chez Panisse

---

§ "Manifesto on the Future of Food" only
† "Manifesto on the Future of Seed" only

BELLY OF THE BEAST $5.99/LB:
THE FUTURE OF FOOD IN THE US

A View from Behind the Counter

jamey lionette

I AM NOT A SCIENTIST, journalist, or other specialist. I sell food. I help run a family-owned and operated neighborhood market and café that buys and sells predominantly local, clean, and sustainable food. I cannot speak about the reality of our food supply around most of the world. I can only can speak of what is happening in the first world, where, unfortunately, only the privileged elite can choose to put real food on their dinner tables.

Lately it seems every mass media newspaper or magazine, from the *New York Times* to *Rolling Stone*, has an article digging into the true filth that most food in the US really is. Some people are actually questioning mass produced and mono-culture organic food. Even *Time* magazine proclaimed "Local Is the New Organic" on its cover. Everywhere I turn people tell me that there is a new wind in the US; that people are now concerned about eating local, clean, and sustainable food.

From my vantage point in the market, behind the counter, I just don't see it. Yes, in Massachusetts there are more farms today than in the last 20 or so years, but fewer total acres than ever recorded. Farmers markets are becoming popular or perhaps trendy. Chain supermarkets are "listening to their customers" and capitalizing on cheap "organic" food. But the chain-supermarket owners are some of the same people who screwed up our food supply in the first place. How can we trust them?

Outdoor food markets are a mainstay in most cultures in the world and were once a given in our culture. Now most people go there to shop for the luxury food treats (locally grown food) and get their staples at the supermarket. I think that because of the Depression (when there was no money to spend on food) and World War II (when there was rationing and everyone was focused on the war effort) Americans lost their taste-buds. Along came the mass-produced foods of the 1950s at cheap prices. Supermarkets were a "progressive" thing, as suburban living was progressive. Rural culture and production was frowned upon as old-fashioned and primitive. Food from all over the world suddenly became available and at prices lower than local food. Protecting America's foreign interest, the beginning of what we now call globalization, became a new form of colonialism. Foreign resources, raw materials as well as labor, were now easily exploitable by the nation's new superpower status. As the economy grew, money filtered down to the managerial and to some of the working class and was coupled with an influx of cheap products made cheaply and available to most classes of the US. Consumerism took off. Our food changed as well, especially with faster transport and technologies trickery to extend the shelf life of food. Seasonal produce became available year round; exotic food (such as bananas and oranges in Boston) became readily available and affordable. Everything was cheaper, the shopping was more convenient, and exotic foods became staples in our diet. Small and local farms shut down or were forced into monoculture farming. A disconnect sprouted between our diets and our food sources. An orange, once a special and rare treat, became an everyday commodity.

Supermarkets are part of mainstream America's identity. Working-class people have little choice but to shop at conventional supermarkets. Middle-class people can shop at places like Whole Foods and appease their consciences with the notion that that food is safer and tastier than conventional supermarket food. And those of the flat earth society—middle- and upper-class people who do not believe that their climate is changing, that a global market is a bad thing, or that our food systems are in trouble—favor the conventional supermarket. However, both conventional and progressive supermarkets operate on the same model: mass-produced foods, made cheaply, and sold at cheap prices.

Supermarkets sell commodities. They buy mass-produced food from big business. This model of efficiency, which mirrored the production of things like automobiles and VCRs, is what created the mess our food supply is in. Efficient ordering and deliveries, no seasonal variety of stock, little to no blemishes (whether natural or from human error), significant quantities—enough to keep all those shelves constantly filled with whatever the customer might want. I describe this model as "I want what I want when I want it," and it goes against everything about food that is local, clean, and sustainable. It cannot be done at a mass level.

Local, clean, sustainable food cannot be mass produced. Let's look at Mr. Clark, a pig farmer. Mr. Clark only has a USDA inspector on Tuesdays and Fridays, and I buy a pig from him every Thursday. Because we have a direct relationship, when I ask he is kind enough to break down our pig into roughly a dozen "primal" pieces (which is very nice because as you can imagine it's a pain in the a** to lug around a 300-pound pig in the kitchen). If you are thinking of preparing a nice pork tenderloin dinner, then come in and ask for it. But if you need three or four we have a problem. Pigs only have two tenderloins. I would need a couple of days' notice to get several pork tenderloins from other local, clean sources who might have them available. I know it seems much easier to just go to a chain supermarket and buy those tenderloins, because they have hundreds of them on hand. After all, efficiency is the key to mass production; having what you want when you want—and the convenience, inexpensive prices, and one-stop shopping—is what puts the "super" before the "market."

However, the manufacturing of this type of food comes at a great cost, even at supermarkets claiming to sell food that is "naturally raised." Supermarkets depend on mass-scale production, transportation, storage, and a huge carbon footprint for their food. You can forget about the markets intimately knowing suppliers. After all, how could a regional manager at a supermarket deal with Mr. Clark when he or she needs 1,000 pork tenderloins? For starters, Mr. Clark doesn't have 500 pigs, and even if he did, he certainly wouldn't slaughter them all just to get 1,000 fourteen- to sixteen-ounce cuts of pork. What would he do with the other 300,000 pounds of pork? And how much would he and his family need to change their farming practices to fulfill such an order? Clark's Farm would have to morph into a mass-produced pork manufacturing facility. It's not unheard of—Stonyfield Farms and Organic Cow of Vermont did just that: neither are in New England anymore (Stonyfield buys some of its milk from local dairy farmers). Wolf's Neck Farm up in Maine did it, too. It is marketed as a small farm in Maine (which it once was), but now Wolf's Neck beef could come from a slaughterhouse in Nebraska or Pennsylvania and not necessarily from New England.

How local is that?

Most restaurants are no better. Margins are so thin that integrity of the food is the first thing to get slashed. There is always a certain amount of labor needed; there is rent, electricity, and other fixed costs that do not, as the name would suggest, have the ability to decrease, so basically cheaper food is a way of turning a profit. Add to this the need for lots of one product such as a certain cut of beef or a particular vegetable, and many restaurant owners turn to mass-produced food which is always available at any quantity at any time of year. Local farms grow seasonally, and weather plays a crucial role in how long and when crops are available. Mass-produced food allows the restaurant owner to think about things other than whether peppers are in season and currently available from local farms or whether a farmer can supply 100 Hanger steaks a week.

Ultimately, the cheap cost of mass-produced food, transported from all over the world is what makes it so attractive to almost every restaurant and supermarket. We are so used to the cheap price of mass-produced food that when we come across real food, it seems extremely expensive. Some people are near outrage when they see the price of our farm-raised food. Far fewer people get outraged that no one in the working class could ever afford to eat real, local, clean, sustainable food. Honestly, would you go to the cheapest doctor you could find—the one with a "Sale! Medical Examinations— Pay what you weigh! Only $1.99/lb!" sign in the window—or the one who would keep you healthy? Americans shop for food like we would for trash bags or broom handles—find the cheapest, use it, and move on. But we are talking about food here. Sustenance. Things we chew and swallow. Things that go deep inside our bodies and fuel us. Food is an integral part of all cultures, and local food producers and farmers are an integral part of our community.

$$\mathcal{Q}$$

PEOPLE FIRST BOUGHT cheap food because they either did not have enough money or felt like they were beating the system by spending less than they budgeted for food that week. Over time our budgets became based on the price of cheap food, so that now, during the rare moment of seeing real food, the price tag appears exorbitant. Our wages and salaries, our rent and utilities, all are tied to our cheaply priced food.

Many people who can actually afford local, clean, sustainable food buy it only when it is trendy, sold at boutique shops, or for a special occasion. Those from the class which struggles to afford mass-produced food certainly cannot afford the real price of food in the US. One often-overlooked agent of gentrification and, after rent increases, one of the best ways to ruin a neighborhood is by shopping at chain supermarkets. Local neighborhood markets close or survive by becoming convenience stores. Farmers' markets become a trendy place to buy a few novelty items: "Oooh look at this peach. I bought it from a farmer!" Once the small markets are gone, only supermarkets are left. We are so out of touch with the struggle to get food, because of how much cheap food is available in the country, that we do not see a pattern of destruction.

The more we buy mass-produced foods, the more it empowers agro-business and the fewer farms there will be. The more we shop at supermarkets, the fewer neighborhood markets there will be. Already we are almost trapped by agro-business and its sales outlets. Soon, there will be no escape. As it stands right now, only a privileged few can afford real, clean, and sustainable food; soon, even the privileged will have little access to such food. The fewer local farms we have, the more expensive their food becomes and the more difficult it is for local farms to feed the local population. Once the farms are gone, only mass-produced food is left.

Hadley, Massachusetts, is known as having the best asparagus in the world. Though just an hour or so outside of Boston, it is near impossible to find asparagus grown in Hadley in Boston. Futures of the asparagus are sold; mostly to France and Japan, I am told. Instead of a wonderful spring vegetable for a local dish, Hadley asparagus has become a boutique item for other parts of the world. Yet in spring, summer, winter, or fall, asparagus flown in from Peru is half the price of in-season asparagus grown on a family farm in New England. And I must admit it seems a bit shameful to complain about such a situation in the US, when so many peoples around the world local resources have been diverted to produce food for Americans.

The late summer is tomato season in New England. The glory of a local tomato salad on a warm summer night in Boston is something which we can only enjoy a couple of months a year. The flavor of our farmers' tomatoes are spectacular. Especially when bought at a local shop or farmers' market, where we actually speak with the people involved in harvesting and distributing our food, people who are part of our community. These tomatoes were not sprayed with anything; the soil was not ruined by chemicals or monoculture farming. These tomatoes traveled only a few dozen miles and were grown outside, thus using only a little energy and creating little pollution. The farmer, part of our community, was deservedly paid and did not exploit anyone or the land. No one was ripped off during the whole transaction, and the tomatoes were available to everyone in Boston during the late summer months.

Yet the rest of the year we still expect to have fresh tomatoes available, and they are called for in many dishes. Fresh tomatoes are considered year-round staples. There is never any questioning tomatoes in March, their integrity or their source. We have become used to hydroponic tomatoes flown in from Mexico or Holland. Instead of focusing our efforts on bringing in tomatoes year-round to Boston, we should focus making the Northeast corridor able to feed itself now and in the future. At the very least, these factory-grown tomatoes do make our local tomatoes taste even more wonderful. We are so used to the mealy, flavorless (or artificially flavored) hydro-tomato that when we taste a real one, it seems so special. This is one reason why local farmers are not perceived as the people who raise our food, but as the producers of specialty items.

Another reason farmers are considered purveyors of specialty foods is their prices. Let us end the idea right now that local, clean, and sustainable foods result in a high profit for the producer and the retailer—trust me, there is absolutely no money in sustainable food. When food is handled as sustenance—not as a commodity—there is little profit to be had. That is why real food is so rare and so hard to come by now. The perverted twist is that it would seem logical that food transported for days around the world would cost more than something fresh and local. But quite the opposite is true. Nobody considers what the true price of real food is. Nobody is outraged that what most working-class people can afford, and even the middle class can afford, is nothing more than mass-produced, cheapened food.

There are, of course, the Whole Foods, Wal-Marts, Trader Joes, and other chain supermarkets trying to sell organic foods. Everyone knows these places are cheaper than local markets and farmers' markets, but rarely do people think about how supermarkets work. People are generally aware of the smaller mark-up chain supermarkets can afford, as compared with an independent neighborhood market, as well as all the corporate capital and funding behind them. But few often think about what is involved in producing enough of a particular food for every shelf of their hundreds or thousands of outlets across the region or country. You can't see the devastating effects of monoculture farming in the sterile and lifeless supermarket. The food looks so perfect and seems so abundant. And with such cheap prices, why ask questions? Sustainable farming does not have the ability to be mass produced; it cannot be sold at the level of a chain supermarkets. Corners must be cut to keep costs low, production must increase to fill the shelves, the laws of nature must be beaten by science to allow for year round production, and if the weather cannot yet be defeated, then the product should be mass-produced and imported from another part of the world.

℘

LISTEN, THANKSGIVING 2006: Whole Foods Boston was selling a "fully pastured naturally raised" turkey for $1.99/lb. That is painfully cheap. Was it trying to compete with the half-dozen small town turkey farmers still left in Massachusetts or the handful of farmers selling turkeys to their regular customers at the farmers' markets or through community-supported agriculture (CSA)? Probably not. Such consumers of locally raised food still have an appreciation for the tradition of buying a turkey from the same place every year or still get pleasure from buying their turkeys directly from their friend, the farmer, or a neighborhood shop. Whole Foods was trying to compete with the other big supermarkets, who sell cheap food.

Whole Foods (and the supermarkets imitating it) will be the death of the movement for clean, local, and fair food for many reasons, but this is an important one. By dropping the price so low, and using claims and slogans designed not by farmers but by slick salespeople, it has set the expectation that clean food can be as cheap as, or just slightly more expensive than, filthy food. Many people could afford to make the jump from Butterball to a Whole Food bird and, with that jump, assume that the bird was safer, more sustainable, and cleaner. So now any farmer charging a real price is seen as greedy or overpriced. Like Wal-Mart's cheap organic, Whole Foods has cheapened (in integrity, as well as price) naturally raised meats and clean food. It lowers the bar by allowing cheap mass-production and corner-cutting, all to sell cheap food that you think is something it is not. There is tokenistic buying of local food and various labels to suggest a certain quality to the consumer. Because we have so few local farms left, it is easy for a chain supermarket to buy some local food and appear to be supportive of local farms. For most people, this is the easy and convenient way to feel as though they are doing the right thing. But it was the supermarket in the first place that helped reduce the number of farms and transformed our understanding of what local farms are.

Organic food is by no means synonymous with clean food. What should we expect, considering a food supply which is mass-produced will be shipped all over the world? And how did the E. coli get into the spinach? Nobody knows. The apparatus is too big. We are concerned, but we are overwhelmed and more importantly completely removed from our food; we have no idea how to eat locally. I am sure nearly half of Boston goes months without ever eating a single bite of local food.

Are people buying store-brand organics duped or misled? Not exactly. The argument for mass-produced organic food is that at least it is a lesser of two evils. I would agree that mass-produced organic or mass-produced naturally raised is not as bad as mass-produced conventional food, but it is still bad. Are we content with eating bad food? Where is the outrage at choosing between bad and worse? Within the first world, on a day-to-day basis, there is barely a struggle to obtain food. But obtaining clean food is a struggle. And to complicate matters are savvy marketing and confusing legal and nonlegal claims. Do the research on what the USDA allows for the claim "free-range"or "organic." They are by no means what you would expect. To be labeled free-range, the law states only that once a bird is old enough to safely venture outside (fair enough, small chicks are at risk outside to predators, weather, diseases, etc.) that they can be kept inside as long as they have access to the outdoors. Often this means a small hole in the wall leading to a small, lifeless patch of land, which the bird never bothers going out to. And for organic—just a few hours outdoors (not necessarily free of a cage) and nothing but USDA certified organic feed. Great, but that feed may not be what that animal wants to eat at all. Mass-produced food and monoculture farming does nothing good for the land. It burns it up. It is not sustainable. Organic or conventional— if it is produced in favor of profit over sustainability it cannot last forever.

More and more, farmers are not paying for the organic certification. Some are too small and unable to satisfy the regulations; others just cannot or do not want to pay for the certification. Why waste the money? Why spend money on a certification which would only bring the farmer down to the same level as industrial mass-produced food sold at a fraction of the farmer's asking price? It's hard to trust big producers and chain stores, but trust is an important part of a healthy community. By buying directly through a farmer or from a neighborhood shop that buys local food, a healthy relationship based on trust can easily grow. When there is a simple and direct connection between the consumer and producer, trust is quite easy.

How many farmers have input with the USDA, and how much input does agro-business have? And how much energy and time does the average working- or middle-class person have to sift through all the marketing terminology and misleading government claims? Where can someone who eats every day actually find clean food? As farms, those places where real food is grown and harvested, verge on extinction, as farmers' markets are primarily accessible and affordable to people of economic privilege, what option is there for most people for clean food? There is now the demand, but the true supply is quite limited, and the profiteers like Wal-Mart and Whole Foods are quick to provide knock-off products. Their success creates even larger challenges for those farming with traditional sustainable models, which cannot compete with mass-production and cheap-food producers.

In the summer and fall, farmers' markets can be found all over the city. No one stops eating in the winter, when all the farmers' markets close. A hundred years ago, what happened in the cities from December through April? Did people stop eating? The same food supply that is emptying our oceans, poisoning our rivers and our bodies, putting farms out of business, and using up a ridiculous amount of energy is the only food supply we know how to subsist on. Through technology and science we don't have to eat seasonally anymore, and exotic foods are possible year-round. But so is nuclear war.

When we fully realize or finally admit the true effects of climate change, peak oil, and globalized food as our primary source of food, we will have to change our relationship with food. Though it may be inconvenient to those whose lives are based on convenience, it will not be the end of the world. We have only our own history to look at to seek the solutions.

The truth is, we can—and historically did—eat locally, all the time, back when people knew how to pickle, preserve, and cure in their own homes. Clearly we're less kitchen savvy than our ancestors, but does that mean we ought give up hope and just buy whatever we feel like at the supermarkets? It's certainly tempting. But not necessarily wise. If you really need to eat oranges every day and you live in Boston, then move to Florida. Otherwise, you will have to adapt your diet to what New England has to offer.

Presently, there is little excuse in the summer and fall not to have three meals a day based primarily on locally produced foods. But even in the late winter it's important to still try to keep it local, at the very least keep it clean, and certainly keep it sustainable. Part of that can be accomplished not only by changing what we eat but also how we buy. If there is no seafood caught sustainably that day, eat beef from a local farm. We can no longer get what we want when we want it. We can get what is available and both enjoy and subsist on it.

I have heard the argument that "organic" or clean or real food is not sustainable because there are not enough farmers and not enough land to produce enough clean food for everyone in the country, let alone the world. Good, clean, and fair food is not sustainable when food is considered a commodity. Good, clean, and fair food is sustainable in a normal society. We must remember, as Michael Pollan writes, "what it means to not be sustainable. If it is not sustainable, then sooner or later it is going to collapse." And what of the rest of the world, who do not have the option of clean or dirty food? What of people whose seeds are being patented by foreign companies so that farmers cannot even afford to grow their food for their communities? Yet, here in the first world, whose problems certainly pale in comparison to the rest of the world, we still must ask, "Is it safe to eat the spinach?" Well, if a first-world nation cannot even manage to feed its people without getting them sick and ruining the land and the future, then good riddance to bad rubbish. As for the spinach, if it is in season, and it is local, then yes, it is safe. If it is out of season, coming from far away, then no, it as dangerous as it was before, during, and after the E. coli problem. Either eat seasonally or be part of a society which runs the risk of destroying itself.

ℒ

THIS IS OUR SOCIETY. A society that has no interest in banning feedlots or the excessive/exclusive feeding of grains, hormones, animal by-products, and antibiotics to cows and seemingly covers up any connection with these practices to E. coli. Worse, our health officials and beef industry leaders come up with a chemical injection to kill possible E. coli and dabble in using pro-biotic injections to make our food "safe." What did you expect? These are the same people who actually believed that forcing cows to be cannibals in confined quarters—which gave us mad-cow disease—for the sake of cheap beef and high profits was not a bad idea. If you could witness how most of our food is produced, you would not eat it; you would be outraged. We are so far removed from our food.

People think that by washing the vegetable with water that all the chemicals are washed off. Even more absurd, many of these same people will buy bottled water because they don't trust the tap water to drink (but they think it is clean enough to rinse their food with?). People don't worry about chemicals possibly absorbed into the food and seeping into the land. People choose shiny fruit covered in wax and pesticide over the uglier, mishapen, dull-colored clean fruit from a farm because they believe it will taste better or is safer. How ludicrous is it when mass-produced food is just called "tomato" or "beef," but real food must be called "NOFA Certified Organic-locally grown on a small, clean, sustainable farm, free of all pesticides heirloom tomato" or "100 percent grass-fed/grass finished, hormone-free, antibiotic-free, animal-by-product–free, fully pastured, naturally raised on a small, local, sustainable family-farm beef." This is a society that has organic corn syrup! There is fair reason to be disgusted and outraged at our current food supply and culture of convenience that has created and perpetuated this mess.

It is nice to believe that eating is a revolutionary act, but sooner or later someone is going to have to call this system out. When a few people start ruining our food, we must take action against those people. When a system has failed, we must change that system. When we are perpetuating that system because of our laziness and lust for convenience, then we must change, or else we will collapse. I cannot think of any point in history when a food supply has been so dangerous. Food's place in our culture and community has faded into cheap traditions. Our planet's fertile land has decayed, been poisoned, and been transformed into factories while we have been too busy and out of touch with our food to notice. The people who know how to use the land to produce food have lost their place on the land, and we did not notice because we no longer know who produces our food. Our food supply is being linked to long-term damage such as heart disease and cancer. And now our food is killing us instantly. Not a week passes it seems that there is not some kind of deadly outbreak. What are you doing about it? We can easily envision a society based on sustainable food; most cultures throughout history have had sustainable farming practices. Basically, Grandma had it right and the progressive supermarkets had it all wrong. We do not necessarily have to turn back the clock and return to an agrarian society, but let's understand what Grandma was doing and realize that she was a lot smarter than we are today. She may not understand the complexities of the internet, but we are the fools who cannot even preserve our summer vegetables so we don't starve in the winter.

We must address the classic American attitude of individuality. Our culture, probably more than any other culture in the world, is based on the individual. Our economic system fuels this individuality. Look at our eating habits. Rather than supporting our community, we buy cheap food from far-away places in chain supermarkets. We do not realize what we are doing to our own community, because we no longer think about our community—we think only of ourselves. Eating can no longer be an individual act. It is not about whether an individual wants to get fat or die from gluttony.

Antibiotics are becoming less and less efficient as pathogens and virus mutate. It seems clear that this is directly related to the excessive use of antibiotics in our food supply. Roughly 75 percent of all antibiotics in this country are given to our livestock. Again, I am not a scientist, but it seems quite clear that even people who only eat antibiotic-free meats will find their medicine useless, as a mutated virus will resist antibiotic treatment regardless of what kind of meat was eaten. The use of pesticides can be equally harmful to the strict organic eater, as a personal choice at the dinner table can do nothing to stop the chemicals of conventional farms from seeping into the rivers and soil. We should all have a right to eat clean, healthy, and sustainable food. It should be a privilege to eat exotic and out-of-season food. Right now, however, we have the right to eat exotic and out-of-season food, and the privileged few can eat clean, healthy, and sustainable food.

When we fully realize or finally admit the effects of climate change, peak oil, and globalized food as our primary source of food, food from international sources will be more expensive than local food. How do we get back to where local food is normal and affordable, and food from far away is exotic and truly expensive? We have successfully wiped out most of the farms and do not have many farmers left. I can only hope that we can start supporting our local farmers—real support, not the tokenistic once in a while local treat. We must face the reality that urban sprawl must give way to farmland. We must realize that we cannot eat beef every day, but, at least when we do it won't kill us. This will involve spending more of our money, but soon the amount we spend on food will feel normal and not expensive. Americans pay less per capita than anyone else in the world for food.

It should be really easy for privileged people to buy fewer luxury items and spend the same percentage of income as other people in the world do on food, but the same cannot be said for the majority of people in the US. Most people in this country are dependent on their weekly wages and live paycheck to paycheck. Wages are set to allow people to survive so they can show up to work. There is little extra money put into that equation for clean, sustainable food.

We could hope that more farms will appear and there will be more farmers to provide enough real food for everyone at an affordable price. We could hope that supermarkets and agro-business would just take care of the problem for us and magically make good, clean, fair, sustainable food cheap enough to fit into our current model. Or hope that these same businesspeople who have ruined our food supply and who are wrecking our land will take their millions of dollars of profit and happily give it back to the farmers and small producers—people who see food as sustenance, not commodity. But that just is not going to happen.

As our food entered our economic systems it was transformed from sustenance to commodity, and I do not see how we can take it back while maintaining this economic system. We have to ask ourselves what we want, food or our current economic system. We need to realize that our system itself is not sustainable and has failed.

℔

YOU HAVE TO BE a little shameless to look at a mother or father and ask what her or his child is going to say upon growing up and there are no more cod in the ocean, no more farmed food, and even less land to farm. What will your child think of you? Maybe he or she will say, "You did nothing, just perpetuated this problem and left it with me—thanks a lot." Shameless? Perhaps, but we need to start thinking this way. We look twice at a parent smoking or getting drunk in front of a child—some would say intervention is necessary. So should I intervene when a privileged mother who can actually afford clean, local raspberries that cost $5.95/lb puts them back in horror of the price and picks up some filthy, mass-produced, conventional black raspberries from another part of the world because they cost only $2.99/lb? Especially when the mother's response is, "They are just for my kids, they won't know the difference." We banned cigarettes from public places in Boston because (after working our way through all the misinformation promoted by the cigarette companies) we realized it was harmful to other people. We decided if someone wanted to smoke, then fine—they can harm themselves, but they are not allowed to smoke where they can cause harm to others. Can we apply this to our food supply? We ban certain harmful products and regulate food production standards for safety issues. But will we regulate based on sustainable criteria? Which takes precedent, our individual right to decide what we eat or sustainability?

As agro-business rises and farms disappear, we have to wonder how we can manage to actually get enough local, clean, and sustainable food for our population. Will everyone have the right to eat it? Will everyone be able to afford to eat it? If we run out of farms and farmers, will there be enough, if any, clean food? When we are out of our local farms, we are out food. At present our whole infrastructure and distribution of local food is in shambles. Believe me, I know it—local farms are where we buy most of our food for our market and café. We must remember that, at present, only the privileged can regularly consume clean, local, and sustainable food. Cheap food also has the hidden cost of weakening communities as well as devastating the land, our bodies, and our future food. Cheap food has no long-term potential; it is only a momentary solution for high profit.

With peak oil and climate catastrophe looming, local and clean food becomes more than just romanticized food culture; it becomes essential. Will our future food supply of local, clean, and sustainable food be possible? It has to be. Is it really so difficult for our culture to return to traditional habits of eating, the habits that have worked for several thousand years? Eat what is local, grow it in a way that does not destroy the land or our bodies, and make it fair to both the people raising the food and eating the food.

You Are What You Grow

michael pollan

A FEW YEARS AGO, an obesity researcher at the University of Washington named Adam Drewnowski ventured into the supermarket to solve a mystery. He wanted to figure out why it is that the most reliable predictor of obesity in the US today is a person's wealth. For most of history, after all, the poor have typically suffered from a shortage of calories, not a surfeit. So how is it that today the people with the least amount of money to spend on food are the ones most likely to be overweight?

Drewnowski gave himself a hypothetical dollar to spend, using it to purchase as many calories as he possibly could. He discovered that he could buy the most calories per dollar in the middle aisles of the supermarket, among the towering canyons of processed food and soft drinks. (In the typical American supermarket, the fresh foods—dairy, meat, fish and produce—line the perimeter walls, while the imperishable packaged goods dominate the center.) Drewnowski found that a dollar could buy 1,200 calories of cookies or potato chips but only 250 calories of carrots. Looking for something to wash down those chips, he discovered that his dollar bought 875 calories of soda but only 170 calories of orange juice.

As a rule, processed foods are more "energy dense" than fresh foods: they contain less water and fiber but more added fat and sugar, which makes them both less filling and more fattening. These particular calories also happen to be the least healthful ones in the marketplace, which is why we call the foods that contain them "junk." Drewnowski concluded that the rules of the food game in the US are organized in such a way that if you are eating on a budget, the most rational economic strategy is to eat badly—and get fat.[*]

This perverse state of affairs is not, as you might think, the inevitable result of the free market. Compared with a bunch of carrots, a package of Twinkies, to take one iconic processed foodlike substance as an example, is a highly complicated, high-tech piece of manufacture, involving no fewer than 39 ingredients, many themselves elaborately manufactured, as well as the packaging and a hefty marketing budget. So how can the supermarket possibly sell a pair of these synthetic cream-filled pseudocakes for less than a bunch of roots?

---

[*] Drewnowski, A. "Obesity and the food environment: dietary energy density and diet costs." *American Journal of Preventive Medicine* 2004; 27(3S):154–162. Reprinted in: *An Economic Analysis of Eating and Physical Activity Behaviors: Exploring Effective Strategies to Combat Obesity* (eds: J.O. Hill, R. Sturm, C.T. Orleans).

For the answer, you need look no farther than the farm bill. This resolutely unglamorous and head-hurtingly complicated piece of legislation, which comes around roughly every five years, sets the rules for the American food system—indeed, to a considerable extent, for the world's food system. Among other things, it determines which crops will be subsidized and which will not, and in the case of the carrot and the Twinkie, the farm bill as currently written offers a lot more support to the cake than to the root. Like most processed foods, the Twinkie is basically a clever arrangement of carbohydrates and fats teased out of corn, soybeans, and wheat—three of the five commodity crops that the farm bill supports, to the tune of some $25 billion a year. (Rice and cotton are the others.) For the last several decades—indeed, for about as long as the American waistline has been ballooning—US agricultural policy has been designed in such a way as to promote the overproduction of these five commodities, especially corn and soy.

That's because the current farm bill helps commodity farmers by cutting them a check based on how many bushels they can grow, rather than, say, by supporting prices and limiting production, as farm bills once did. The result? A food system awash in added sugars (derived from corn) and added fats (derived mainly from soy), as well as dirt-cheap meat and milk (derived from both). By comparison, the farm bill does almost nothing to support farmers growing fresh produce. A result of these policy choices is on stark display in your supermarket, where the real price of fruits and vegetables between 1985 and 2000 increased by nearly 40 percent while the real price of soft drinks (aka liquid corn) declined by 23 percent. The reason the least healthful calories in the supermarket are the cheapest is that those are the ones the farm bill encourages farmers to grow.

A public-health researcher from Mars might legitimately wonder why a nation faced with what its surgeon general has called "an epidemic" of obesity would at the same time be in the business of subsidizing the production of high-fructose corn syrup. But such is the perversity of the farm bill: the nation's agricultural policies operate at cross-purposes with its public-health objectives. And the subsidies are only part of the problem. The farm bill helps determine what sort of food your children will have for lunch in school tomorrow. The school-lunch program began at a time when the public-health problem of America's children was undernourishment, so feeding surplus agricultural commodities to kids seemed like a win-win strategy. Today the problem is overnutrition, but a school lunch lady trying to prepare healthful fresh food is apt to get dinged by USDA inspectors for failing to serve enough calories; if she dishes up a lunch that includes chicken nuggets and Tater Tots, however, the inspector smiles and the reimbursements flow. The farm bill essentially treats our children as a human Disposall for all the unhealthful calories that the farm bill has encouraged American farmers to overproduce.

To speak of the farm bill's influence on the American food system does not begin to describe its full impact—on the environment, on global poverty, even on immigration. By making it possible for American farmers to sell their crops abroad for considerably less than it costs to grow them, the farm bill helps determine the price of corn in Mexico and the price of cotton in Nigeria and therefore whether farmers in those places will survive or be forced off the land, to migrate to the cities—or to the US. The flow of immigrants north from Mexico since the North American Free Trade Agreement (NAFTA) is inextricably linked to the flow of American corn in the opposite direction, a flood of subsidized grain that the Mexican government estimates has thrown 2 million Mexican farmers and other agricultural workers off the land since the mid-90s. (More recently, the ethanol boom has led to a spike in corn prices that has left that country reeling from soaring tortilla prices; linking its corn economy to ours has been an unalloyed disaster for Mexico's eaters as well as its farmers.) You can't fully comprehend the pressures driving immigration without comprehending what US agricultural policy is doing to rural agriculture in Mexico.

And though we don't ordinarily think of the farm bill in these terms, few pieces of legislation have as profound an impact on the US landscape and environment. Americans may tell themselves they don't have a national land-use policy, that the market by and large decides what happens on private property in the US, but that's not exactly true. The smorgasbord of incentives and disincentives built into the farm bill helps decide what happens on nearly half of the private land in the US: whether it will be farmed or left wild, whether it will be managed to maximize productivity (and therefore doused with chemicals) or to promote environmental stewardship. The health of US soil, the purity of its water, the biodiversity and the very look of its landscape owe in no small part to impenetrable titles, programs, and formulae buried deep in the farm bill.

Given all this, you would think the farm-bill debate would engage the nation's political passions every five years, but that hasn't been the case. If the quintennial antidrama of the "farm-bill debate" holds true to form this year, a handful of farm-state legislators will thrash out the mind-numbing details behind closed doors, with virtually nobody else, either in Congress or in the media, paying much attention. Why? Because most of us assume that, true to its name, the farm bill is about "farming," an increasingly quaint activity that involves no one we know and in which few of us think we have a stake. This leaves our own representatives free to ignore the farm bill, to treat it as a parochial piece of legislation affecting a handful of their Midwestern colleagues. Since we aren't paying attention, they pay no political price for trading, or even selling, their farm-bill votes. The fact that the bill is deeply encrusted with incomprehensible jargon and prehensile programs dating back to the 1930s makes it almost impossible for the average legislator to understand the bill should he or she try to, much less the average citizen. It's doubtful this is an accident.

But there are signs this year will be different. The public-health community has come to recognize it can't hope to address obesity and diabetes without addressing the farm bill. The environmental community recognizes that as long as we have a farm bill that promotes chemical and feedlot agriculture, clean water will remain a pipe dream. The development community has woken up to the fact that global poverty can't be fought without confronting the ways the farm bill depresses world crop prices. They got a boost from a 2004 ruling by the World Trade Organization that US cotton subsidies are illegal; most observers think that challenges to similar subsidies for corn, soy, wheat, or rice would also prevail.

And then there are the eaters, people like you and me, increasingly concerned, if not restive, about the quality of the food on offer in the US. A grassroots social movement is gathering around food issues today, and while it is still somewhat inchoate, the manifestations are everywhere: in local efforts to get vending machines out of the schools and to improve school lunch, in local campaigns to fight feedlots and to force food companies to better the lives of animals in agriculture, in the spectacular growth of the market for organic food and the revival of local food systems. In great and growing numbers, people are voting with their forks for a different sort of food system. But as powerful as the food consumer is—it was that consumer, after all, who built a $15 billion organic-food industry and more than doubled the number of farmers' markets in the last few years—voting with our forks can advance reform only so far. It can't, for example, change the fact that the system is rigged to make the most unhealthful calories in the marketplace the only ones the poor can afford. To change that, people will have to vote with their votes as well—which is to say, they will have to wade into the muddy political waters of agricultural policy.

Doing so starts with the recognition that the "farm bill" is a misnomer; in truth, it is a food bill and so needs to be rewritten with the interests of eaters placed first. Yes, there are eaters who think it in their interest that food just be as cheap as possible, no matter how poor the quality. But there are many more who recognize the real cost of artificially cheap food—to their health, to the land, to the animals, to the public purse. At a minimum, these eaters want a bill that aligns agricultural policy with our public-health and environmental values, one with incentives to produce food cleanly, sustainably, and humanely. Eaters want a bill that makes the most healthful calories in the supermarket competitive with the least healthful ones. Eaters want a bill that feeds schoolchildren fresh food from local farms rather than processed surplus commodities from far away. Enlightened eaters also recognize their dependence on farmers, which is why they would support a bill that guarantees the people who raise our food not subsidies but fair prices. Why? Because they prefer to live in a country that can still produce its own food and doesn't hurt the world's farmers by dumping its surplus crops on their markets.

The devil is in the details, no doubt. Simply eliminating support for farmers won't solve these problems; overproduction has afflicted agriculture since long before modern subsidies. It will take some imaginative policy-making to figure out how to encourage farmers to focus on taking care of the land rather than all-out production, on growing real food for eaters rather than industrial raw materials for food processors, and on rebuilding local food economies, which the current farm bill hobbles. But the guiding principle behind an eaters' farm bill could not be more straightforward: it's one that changes the rules of the game so as to promote the quality of our food (and farming) over and above its quantity.

Such changes are radical only by the standards of past farm bills, which have faithfully reflected the priorities of the agribusiness interests that wrote them. One of these years, the eaters of the US are going to demand a place at the table, and we will have the political debate over food policy we need and deserve. This could prove to be that year: the year when the farm bill became a food bill, and the eaters at last had their say.

Here are some key groups working on reforms of the farm bill—many offer ways to sign up for emailed updates.

- ❧ Public Health Action on the Farm Bill:
  www.publichealthaction.org
- ❧ National Campaign for Sustainable Agriculture:
  www.sustainableagriculture.net
- ❧ Environmental Working Group:
  www.ewg.org
- ❧ National Family Farm Coalition:
  www.nffc.net
- ❧ Slow Food USA:
  www.slowfoodusa.org/farmbill
- ❧ Om Organics:
  www.omorganics.org
- ❧ Community Alliance with Family Farmers:
  www.caff.org/policy/2007_farm_bill.shtml
- ❧ Institute for Agriculture and Trade Policy:
  www.agobservatory.org/issue_farmbill2007.cfm
- ❧ Community Food Security Coalition:
  www.foodsecurity.org
- ❧ Farm and Food Policy Project:
  www.farmandfoodproject.org/index.asp
- ❧ American Farmland Trust:
  www.farmland.org/programs/campaign/default.asp
- ❧ Another great resource is the book *Food Fight: The Citizen's Guide to a Food and Farm Bill*, by Dan Imhoff.

Resources

HERE IS A LIST of contact information for organizations involved with drafting the Manifestos. Please get in touch to help build a sustainable future.

- ❧ Food First/Institute for Food and Development Policy
  398 60th Street, Oakland, CA 94618
  www.foodfirst.org/

- ❧ Institute for Agriculture and Trade Policy
  2105 First Avenue South, Minneapolis MN 55404
  www.iatp.org/

- ❧ Slow Food USA
  20 Jay Street, Suite 313, Brooklyn, NY 11201
  www.slowfoodusa.org/

- ❧ International Society on Ecology and Culture
  PO Box 9475, Berkeley, CA 94709
  www.isec.org.uk/

- ❧ Small Planet Institute
  25 Mt. Auburn St., Suite 203, Cambridge, MA 02138
  www.smallplanetinstitute.org/

- International Center for Technology Assessment
  660 Pennsylvania Ave SE, Suite 302,Washington, DC 20003
  www.icta.org

- Sustain: The Alliance for Better Food and Farming
  94 White Lion Street, London N1 9PF United Kingdom
  www.sustainweb.org/

- International Federation of Organic Agricultural
  Movements
  Charles-de-Gaulle-Str. 5, 53113 Bonn, Germany
  www.ifoam.org/

- The Swedish Ecological Farmers Association/Ekologiska
  Lantbrukarna i Sverige, Sågargatan 10 A
  S- 753 18 Uppsala, Sweden
  www.ekolantbruk.se

- Navdanya
  Research Foundation for Technology, Science & Ecology
  A-60, Hauz Khas, New Delhi-110016, India
  www.navdanya.org/

Related Titles